GROW
MORE
FOOD

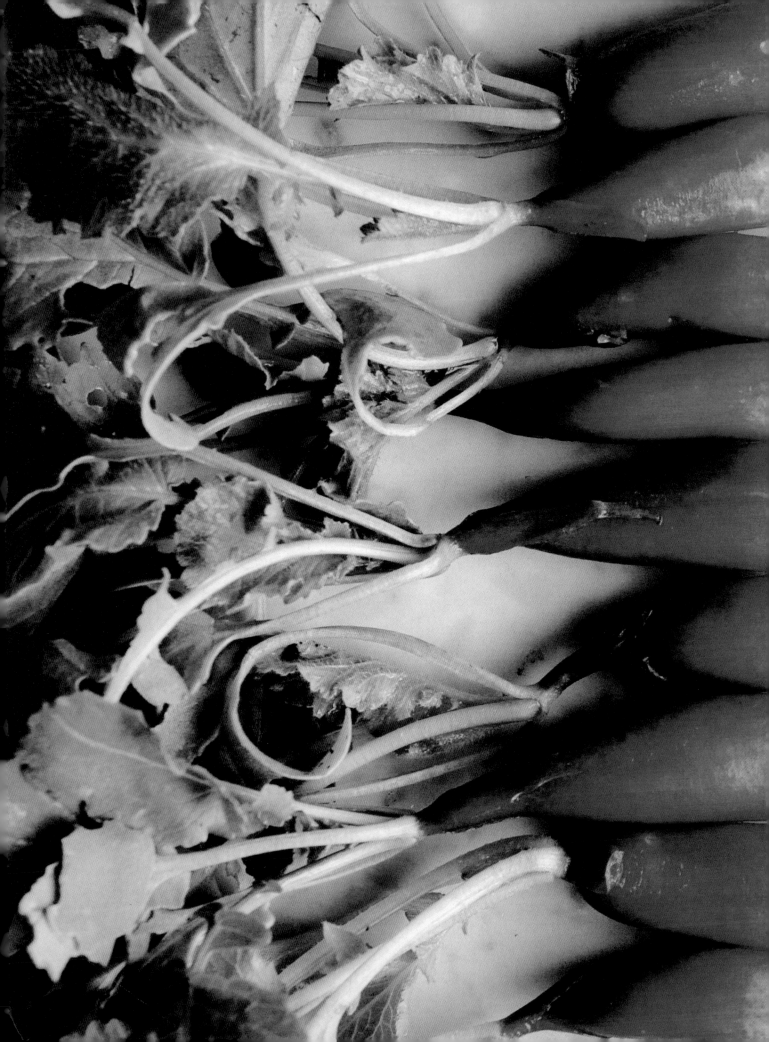

GROW
MORE
FOOD

A Vegetable Gardener's Guide
to Getting the Biggest Harvest
Possible from a Space of Any Size

COLIN McCRATE & BRAD HALM
PHOTOGRAPHY BY HILARY DAHL

Storey Publishing

The mission of Storey Publishing is to serve our customers by publishing practical information that encourages personal independence in harmony with the environment.

Edited by Carleen Madigan
Art direction by Ash Austin
Book design by Stacy Wakefield Forte
Text production by Jennifer Jepson Smith
Indexed by Claire Splan

Cover and interior photography by © Hilary Dahl
Additional photography by © Andrei310/iStock.com,
    197; © Nigel Cattlin/Alamy Stock Photo, 198;
    Courtesy of Waypoint Analytical, LLC, 71
Illustrations by © Steve Sanford, except 48 & 52 by
    Ilona Sherratt © Storey Publishing, LLC

Text © 2022 by Colin McCrate and Brad Halm
A previous version of this book was published under
    the title *High-Yield Vegetable Gardening* (Storey
    Publishing, 2015).

Storey books are available at special discounts when purchased in bulk for premiums and sales promotions as well as for fund-raising or educational use. Special editions or book excerpts can also be created to specification. For details, please call 800-827-8673, or send an email to sales@storey.com.

Storey Publishing
210 MASS MoCA Way
North Adams, MA 01247
storey.com

Printed in the United States by Bradford & Bigelow
10 9 8 7 6 5 4 3 2 1

Library of Congress Cataloging-in-Publication Data
    on file

This book is dedicated to Roy Brubaker, who has inspired many people to grow good food for the world.

CONTENTS

3

GET TO KNOW YOUR PLANTS

Grow more food by planting the right varieties at the right time with the best care.

4

CREATE EFFICIENT SYSTEMS

Plant more of the crops you want & water them efficiently.

5

EXTEND & EXPAND THE HARVEST

Harvest longer & more often & store crops properly.

BECOME A MORE PRODUCTIVE VEGETABLE GARDENER

↓ This book is for people who are intent on getting as much food as possible from their gardens, whatever the size of their plot. Our knowledge is drawn from a background in small- and large-acreage farming, as well as in backyard gardening. After years working on a variety of diversified vegetable farms, we launched a home gardening business. Since 2007, we've been running this business, called Seattle Urban Farm Company, which helps homeowners, restaurants, and communities grow their own food. We've taken the systems and practices used by professional growers and adapted them for use at the scale of a home garden.

There are many reasons why a home gardener might want to implement the strategies described in this book. It may be that you simply want to be more efficient with your time and resources. Or feed your household from the garden year-round. Perhaps you'd like to set up a miniature farm stand by the mailbox for supplemental income, sell a few vegetables at your local farmers' market, coordinate a community garden, or prepare for the zombie apocalypse.

A well-organized garden will have room for dozens of different crops in a small amount of space.

No matter what your goals are, the strategies and techniques described in this book will help you approach your garden like a small-scale farm, dramatically increasing its productivity. Focusing on production means that your plot will need more careful planning, record keeping, and management than a typical backyard vegetable garden.

A vegetable garden can be both beautiful and productive.

HOW TO GROW MORE FOOD IN THE SPACE YOU HAVE

The first step toward a highly productive garden is simply understanding what makes a garden successful. To get the most out of your garden, it's important to:

Grow for a purpose. Take the time to consider the goals of your project. Grow for the tastes you prefer and yields you can use. Make sure to have a use in mind for each crop before it goes in the ground.

Select the best site and use it efficiently. Think ahead and place annual and perennial crops in appropriate locations. For most crops and climates, more sun is better. Lay out your garden to maximize productive space and find creative solutions for spaces outside of the main vegetable garden. Keeping a productive garden space requires using nongarden spaces in support roles.

Plan well and keep good records. Spend time before each season to make a thorough plan of the garden. Update the plan throughout the season as you make necessary changes. Maintain an accurate record of garden tasks and what happens in the garden, and use this information to inform future plans.

Maintain fertile soil. Successful growers say, "Care for the soil, not the crops." Ongoing and meticulous care of the soil is essential. Soil amendment should happen several times every year.

Extend your broccoli harvest by picking side shoots as well as the main crown.

Know your plants. To get the most out of your crops, you must develop an understanding of the physiology, genetics, and cultural requirements of the plants. The more you know about your crops, the easier it is to increase their yields.

Select the best crops. Choose crops and specific cultivars that will perform well in your climate. You'll want to select varieties that are vigorous, produce well, and that you like to eat.

Deal with pests, diseases, and weeds immediately. Closely and frequently monitoring the garden for problems allows you to deal with them before they get out of hand.

Observe and respond. You are the best ongoing source of information about your garden. Keeping track of which varieties perform best and what pests show up and when will enable you to customize your project to your specific conditions.

Maximize your time and energy. Develop systems and use tools that maximize efficiency by saving time and energy. Time is nearly always the most limited resource of a production gardener, so make the most of it.

Water well. Vegetable plants need consistent and adequate water. By the time you notice signs of water stress, you've already reduced your overall yield potential.

Extend and expand the seasons. Create spaces that allow you to extend planting and harvest dates earlier and later in the season. Stay organized with succession planting to grow multiple crops in each space throughout the year.

Harvest and store crops smartly. Know the appropriate time of day and stage of growth to harvest your crops. Pay attention to postharvest care for maximum quality and storage life.

THINK LIKE A FARMER

↓ In our experience, the most successful growers have a decidedly positive way of thinking about their gardens. Although intensive food production is challenging, these growers understand that you are most successful when you find joy in the process itself. They don't allow the inevitable struggles to tarnish their experience.

The authors take a moment to reflect on the day's work.

There is no doubt that intensive food production is hard work and can be exhausting and frustrating. Crops will fail, seasons will be unexpectedly hot or cold, and more insects than you imagined possible will eventually cross your path. One day you may find yourself screaming obscenities and flapping your arms in the air like a deranged chicken in response to a Colorado potato beetle outbreak. However, it's essential to seek creative ways of overcoming these challenges and to find joy in the simple pleasure of doing a little better each season. A successful grower recognizes the highs and lows as part of the agreement to work with nature.

To be successful and to improve their growing systems year after year, gardeners must relish the opportunity to learn from their mistakes and from the vagaries of nature. Observation and knowledge are key in this process: Farmers understand that learning how their crops grow and how the plants respond to their care is vital to their livelihood and well-being. It is embracing this interaction and the give-and-take with nature that makes food production so captivating and rewarding.

THE ART OF GARDENING FOR PRODUCTION

You might say that production gardeners are a bit like artists. Their work runs like a thread through all of their activities; they may be struck by sudden inspiration at any moment; and they perform their tasks as dictated by the work, instead of trying to fit their work into a scheduled routine. The soil is your canvas, plants are your medium, and each onion, pepper, and head of lettuce is a work of art.

Like artists, many growers find that their passion increases over time. As you become intimately familiar with your crops and your soil, your techniques will become second nature and you will truly get lost in your work. The most successful growers are those who continue to find new inspiration in their crops and systems. For some, experimenting with new varieties every year helps keep them engaged and motivated. For others, achieving a continual harvest of salad greens or breeding their own variety of winter squash fuels the fire.

Anyone can become an artist in their garden. No matter the size of your plot, you'll find that as your knowledge and experience grow, so will your yields and your love of food production.

Learning the skills and techniques farmers use can help turn your backyard into a miniature farm.

HOW TO USE THIS BOOK

This is a guide to maximizing garden productivity at home, not a comprehensive encyclopedia of vegetable production. We include useful information for growers at every skill level and focus on techniques that will increase yields for the production-minded gardener. We show you the systems and techniques used on small vegetable production farms every day so that you'll learn to use your yard just like a miniature farm.

We focus on annual vegetables and herbs but also include information on perennials. The information is structured this way because we've found that most home gardeners are concerned primarily with their annual crops. Perennials take less care and often take up less space in the grower's mind.

Developing a system to employ all of the techniques outlined throughout the book will take time. While this book will help improve your garden productivity from day one, you should approach gardening with a long view and gradually acquire skills as they become relevant and appropriate for your garden.

It should be no surprise that professional growers take their work very seriously. They monitor every aspect of their farm: recording when crops are planted, fertilized, irrigated, weeded, thinned, pruned, and harvested. They note which varieties perform best, and they continuously make adjustments to their practices as they develop more efficient and successful ways of caring for their crops.

But don't take it too seriously. Gardening is fraught with challenges. It will make you lose your mind if you don't take a step back every now and then and laugh at the whole process. Get away from your garden, play a game of Parcheesi or Ping-Pong—whatever it takes to calm down after you've lost all your tomatoes to late blight. With time, consideration, and a sense of humor, you will become a successful, highly productive gardener.

PART 1

PLAN AHEAD & KEEP RECORDS

INCREASE YIELDS BY USING SPACE EFFICIENTLY, TIMING YOUR PLANTINGS & PLANTING MORE OFTEN.

Taking the time to prepare a thoughtful plan is one of the most important things you can do for your garden. You'll need to consider which crops to grow, how much of each crop to plant, when to plant, and when to expect your harvest. It's no exaggeration to say that a detailed garden plan alone can double or triple the productivity of a garden.

CHAPTER ONE

EVALUATING YOUR CURRENT GARDEN SETUP

If you already have a garden, take the time to consider any challenges it presents. Seemingly small details can make a big difference. Maybe your beds never get enough compost, because the pathways are so narrow that a wheelbarrow won't fit between them. Maybe it's located near a giant pine tree that sends new roots into your beds every spring. Maybe it's already perfect. In any event, it's a good idea to think holistically about your space and try to identify opportunities for improvement.

Late-summer garden harvests are bountiful and diverse.

what are your growing priorities?

⬇ Why are you interested in increasing the productivity of your garden? Do you want to sell tomatoes to the local corner store? Do you want the most diverse range of edible plants possible? It's important to identify your personal goals so you can lay out the garden properly and dedicate appropriate amounts of space to each crop.

For example, if you want to grow salad for dinner every night, you'll have to determine how much salad your household uses, how many nights per week you eat at home, how long it will take each planting to grow, and how many plantings you'll need to make through the year. This planning process is actually a lot easier than it sounds, and it can be fun to do, as you'll see in Chapter 2.

Write down your big-picture priorities as clearly as you can, and use them as a reference during the design process. You'll need to keep these goals in mind as your guiding light through the process, so you stay on track and actually get what you want from your project.

HOW MUCH GARDEN?

The chart below provides some general guidelines on the needs and potential productivity of different gardens according to size. Please keep in mind that these are rough estimates: Everyone uses a garden differently. Not everyone is growing food for the same number of people, and those people will vary in age and dietary preferences. The time needed to manage your space will also depend on how tidy and weed-free you like to keep it.

Spend some time considering the full range of time commitments you have throughout the year. This will help you home in on which garden size will work best for you. If you're just getting started with your garden, we recommend starting small. It's relatively easy to expand a garden in subsequent years, but it can be overwhelming to manage a large garden if you don't have much experience.

A 90-square-foot garden

A 150-square-foot garden

GARDEN SIZE	MANAGEMENT NEEDS
100–200 square feet	An appropriate size for a beginning gardener who wants to try a few different crops and eat consistently from the yard during peak harvest season. Half an hour to an hour a week will be enough to keep up with all garden tasks.
200–400 square feet	A good size for the intermediate gardener with a hectic schedule. It will yield adequate fresh produce for one to four people throughout spring, summer, and fall, with some produce left for putting up. An hour or two of work a week will be sufficient.
400–800 square feet	A group of two to six people can expect to eat fresh from the garden during the main growing season and also harvest quite a bit for late-fall and winter storage. A space this size will require at least 2 to 3 dedicated hours per week for upkeep, harvesting, and processing of crops.
800–1,500 square feet	Large enough to feed four to eight people through the growing season and produce enough storage vegetables to supplement your diet through much of the winter. Plan to spend at least 4 to 6 hours a week managing the space for maximum production and appearance.

GARDEN SIZE	MANAGEMENT NEEDS
1,500–2,000 square feet	Large enough to feed 6 to 10 people during the season and still distribute small quantities of especially productive crops. With proper planning, it's possible to grow substantial storage crops and cool-weather greens. Expect to spend 6 to 8 hours a week keeping up with the garden.
2,000–4,000 square feet	Entering into the realm of a serious undertaking, with a garden that will supply 8 to 15 people with fresh produce through much of the season. Keeping up with this much space will require at least 8 to 12 hours a week. During peak harvest season, you may need to spend several nights a week processing and storing your crops.
4,000–8,000 square feet	A very substantial home vegetable garden, this much space will feed up to 20 people and may also provide a few crops for wider distribution. Plan to spend 12 to 15 hours a week, plus extra time for processing and distribution, as needed.
8,000–15,000 square feet	The largest home garden we have seen falls in this range. You will have the opportunity to produce great quantities of food year-round for up to 25 people. Plan to spend 15 to 20 hours a week managing your space.
15,000–22,000 square feet	Managing this much space will be a part-time job. Expect to spend at least 20 hours per week or more. A garden this size can feed dozens of people and can provide opportunities for storage, processing, and selling of produce.
22,000–44,000 square feet	This is an endeavor large enough to require a full-time or three-quarter-time manager. This range is approximately half an acre to 1 acre (there are 43,560 square feet per acre). A garden this size is a serious endeavor and will likely require additional equipment and supplies that are beyond the scope of this book. The techniques and systems we describe will be very applicable to your project, but you'll need to research the equipment necessary to effectively manage such a space.

A 1,200-square-foot garden

A 4,000-square-foot garden

CREATE A MAP OF YOUR PROPERTY

If you're starting a garden from scratch, creating a detailed map of your yard will help you think about your space holistically and place the elements of your garden in the best location possible. If you have an existing garden, this process can help you reorganize or expand your garden and find ways to add new production spaces.

You can draw a site plan any time of year, but the off-season is a great time to get started. Knowing that you have a few months of lead time will make you less likely to rush or cut corners, as you might be prone to do when you are eager to get your plants in the ground. If you already have a garden on your property, the best time to create your map is right after the end of a growing season. Putting the pieces together is often easiest when you're cleaning up your summer crops and the past year's successes and failures are fresh in your mind.

CREATE A BASE MAP

To create your site plan, you'll need to start with a base map of the space you'll be working with. Try to make the representation of your property as accurate as possible, so that different areas are drawn in scale to each other.

In order for areas and elements to appear in scale on the map, you'll need to measure how large and how far apart they are in your yard. Also measure the edges of your property line, the dimensions of your house, the location of any other notable items (such as the driveway or walkways), existing planting beds, and the placement of water spigots. Even small details can matter. For example, a dryer vent will blast very warm, drying air outside and can coat nearby plants in a film of lint, both of which can result in stress and lower productivity.

Using online resources will make mapping your property much easier. Depending on your location and other variables, such as how much tree canopy cover is on your property, you might be able to trace an image of your lot in Google Earth and print the image, or even export it to a computer drawing program such as Canva, SketchUp, AutoCAD, or Inkscape.

You may be able to use a few other shortcuts to create your property map. For example, you may have received a map of the property when you purchased your home. If not, you may be able to request a map of your property from the city or town government. If you have an existing map, scan it into a computer or trace it onto a new sheet of paper as a starting point for your garden design. Be sure to draw in elements like utility lines, trees, ornamental landscaping, fences, patios, decks, water sources, streets, alleyways, light poles, and known time capsules.

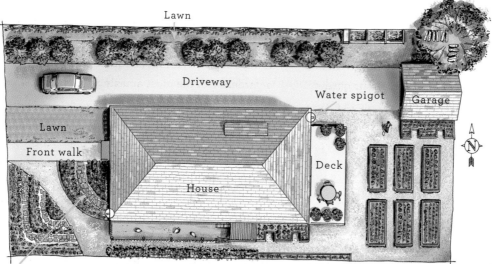

Lawn

Driveway

Water spigot

Garage

Lawn

Front walk

Deck

House

Existing ornamental beds

The first step to planning the garden is to create a base map. This one shows existing structures, impermeable surfaces, and landscape elements.

A sunny front yard is terraced into a productive and easy-to-access garden space.

selecting the best site

Whether you already have a garden or are planning to build a new one, we recommend assessing your site to ensure you're utilizing every advantage your yard has to offer. Poke around your site like an amateur detective. Take a holistic approach and consider if your garden is getting enough sun, if you're making use of beneficial microclimates, and if your garden is laid out in a way that's comfortable and efficient to work in.

FULL SUN = HIGHEST PRODUCTION

Is your garden getting enough sun? Rule number one in the vegetable gardener's handbook is that garden beds should receive a minimum of 6 hours of direct sunlight per day. There is a direct correlation between hours of sunlight and plant productivity, and with a few exceptions, more sun is always better. Without adequate sunlight, even a garden with the most amazing soil and meticulous care will produce leggy, stressed crops and minimal harvests.

There are several new technologies that can help you identify sun exposure, and we encourage you to try them out. If you have a smartphone or other mobile device, try an app called Sun Seeker. This app will show you a three-dimensional view of your space, including the trajectory of the sun on that day, and on the winter and summer solstice. There are probably other similar apps on

WITHOUT ADEQUATE SUNLIGHT, EVEN A GARDEN WITH THE MOST AMAZING SOIL & METICULOUS CARE WILL PRODUCE LEGGY, STRESSED CROPS & MINIMAL HARVESTS.

The south side of a home can trap heat, warming the soil and air to help crops grow more quickly.

the market, so explore the options and find one that works for you. Barring the use of such technology, you can simply make note of the sunlight and shadows on your garden at different times of day. Keep track of the time the sun first hits the garden in the morning and when it falls into shade in the evening. Remember that this span will change throughout the season, so it's a good idea to monitor the sun exposure throughout the year to get a clear picture.

UTILIZE BENEFICIAL MICROCLIMATES

Every property has a unique set of climatic conditions, which contribute to the location's "microclimate." Wind patterns, sun exposure, and the building materials on your property all have significant effects on plant health and productivity. Building materials can have surprisingly large impacts. For example, a concrete driveway will capture and radiate much more heat than a gravel driveway. Similarly, an exposed concrete foundation will capture and radiate more heat than a wall with wooden siding. Keep the following points in mind, as you're identifying microclimates on your site:

+ **Heat sinks.** The exterior walls of your home, garage, shed, or other outbuildings capture heat from the sun and radiate it back out into the environment, so areas on your property that are adjacent to east-, west-, or south-facing walls often have higher temperatures. Ground-level pavement or stonework can also act as a heat sink—your driveway or patio might be several degrees warmer than a nearby lawn. Heat sink microclimates are suitable for tomatoes, peppers, eggplant, melons, and basil.

+ **Wind breaks.** Hedges and fences can help disrupt prevailing winds. Tall vining plants like pole beans, cucumbers, and tomatoes are susceptible to wind damage, as are shallow-rooted crops like peppers. Leafy salad greens also benefit from wind breaks, which help reduce moisture loss from their tender leaves.

+ **Afternoon shade.** While we typically say that more sun is better, in hot climates, heat-sensitive crops desperately need respite from intense afternoon heat. Even in more mild regions, crops like lettuce and spinach can benefit from a midday break.

+ **Dry spots.** The areas underneath trees and the eaves of buildings may have extremely dry soil that is not suitable for growing most vegetables.

+ **Cold spots.** Areas at the base of a hill may be cooler than nearby areas. Just as hot air rises, cool air sinks, and cool air pockets can settle at the bottom of a hill. Properties or portions of a property that lie in a valley may experience earlier and later frosts each season than the surrounding area.

THE "HEAT ISLAND" EFFECT

Urban areas are typically warmer than rural areas so, as a general rule, the closer to a city you live, the warmer your property will be. Man-made structures tend to absorb heat, so the density of buildings and pavement correlates to a warmer local environment. This is often referred to as the "heat island" effect.

DEEP, RICH SOIL WITH LITTLE GRADE

In a perfect world, the soil in your garden will be deep, rich, and full of organic matter, primed and ready for vegetable growing. From our experience, this is a very uncommon occurrence. Most likely, you will be spending some time and effort improving the soil on-site (see Part 2).

Don't worry too much if the soil in your garden needs help. If you're concerned about lead or arsenic contamination, get a professional soil test (see page 69). Otherwise, consider soil improvement as an ongoing part of your maintenance practices and don't rule out a space for new beds just because it seems to have poor soil.

Does your garden have a grade change? A very slight slope of 3 percent or less is easily managed in a garden, but greater angles will result in soil erosion, poor water infiltration, and uncomfortable working spaces. Plan to terrace or regrade garden areas as necessary, or focus on building new beds in flatter locations, if possible.

EFFICIENCY & ACCESSIBILITY

Standardize your beds. We recommend that you plan to create planting beds no wider than 4 feet, so that you can always reach into them without stepping into the bed. Wider beds will make management challenging or create dead spaces in the beds that you simply can't reach. If possible, use standardized bed sizes throughout your entire garden. This allows you to reuse materials in different beds without any modification. For example, if all of your garden beds are 4 by 10 feet, then you can reuse a piece of row cover, bird netting, or irrigation tubing on any bed from season to season.

Create permanent, mulched paths. Make sure you include enough pathways to make movement through the garden easy. Paths can be any width that is comfortable—use 2 feet wide as a starting point and adjust up or down to meet your needs. For larger gardens, consider including a wider access path for wheelbarrows or garden carts. In most locations, for efficiency of space and ease of use, we prefer 4-foot-wide rectangular beds with 2-foot-wide pathways mulched with newspaper and straw, wood chips, or bark mulch to keep down the weeds.

Ensure access. Not every garden can be placed right outside the back door, but you'll want to make accessing the garden as quick and simple as possible. All those wasted minutes running all the way around the house to grab a tool or packet of seeds add up during the course of a season. Even small inefficiencies, repeated over and over, can become obstacles to effective garden care. You'll also want to consider access to water when locating your garden beds. No matter what irrigation system you employ, it is essential that watering the garden not be a hassle or an additional burden on your time. Identify all of your available water lines and determine the easiest routes to bring that water to your garden site.

WHY GROW IN BEDS?

The old way of planting a home garden was to till up a squarish patch of soil (say, 20 by 20 feet), spread a bunch of compost and organic fertilizer, and plant all your crops in rows in that big patch. At the end of the season, you'd clear everything out, and the following spring, repeat the process.

One problem with this method is that when you walk between two rows, you'll probably be walking where you will be planting in the following year. Walking on the soil compacts it, crushing its structure and restricting the movement of air and water through the soil ecosystem. The result is a decrease in microbiological life, poor root development, and ultimately stunted or dying plants. If you compact the soil this way, you'll need to spend extra time loosening that soil again prior to your next planting.

Another downside to growing in rows is that when you add soil amendments over the entire area, you end up spreading them both in your planting spaces (where they're needed) and in your walking paths (where they won't do any good). You'll spend as much time fertilizing weeds as you do fertilizing crops.

We recommend dividing up your garden space into separate, permanent beds and permanent pathways. Keeping pathways untilled and mulched will keep latent weed seeds suppressed and reduce your overall workload. You'll add soil amendments only to the spaces where you actually need them. The bed system also makes it easier to develop a crop rotation plan (see page 49) and to conceptualize planting by the square foot. We still like to plant direct-seeded crops in rows, but we evenly space the rows across the bed as appropriate for the crop. For example, we might plant seven rows of carrots spaced 6 inches apart across the top of a 4-foot-wide bed, or three rows of beans spaced 18 inches apart.

Raised beds allow you to import high-quality soil, keep the space organized, and create weed-suppressing pathways.

ADDITIONAL GARDEN ELEMENTS

Following are a few elements that play an important role in a production garden:

Fencing. Pests will literally take a bite out of your productivity! If deer, rabbits, unruly dogs, or other large animals are a problem in your garden, consider fencing. The size and type of fence will depend on the specific concern, but large animal intrusion is a big enough threat that it should be addressed early on in your planning process.

Propagation and nursery area. Propagating and growing out your own plants is a fun and money-saving process that extends the gardening season, allows you to manage your garden more intensively, and connects you more closely to the garden. Propagating your own transplants is not absolutely necessary, but it is very helpful if you have an appropriate space for it.

Composting area. Highly productive gardens produce more food and more plant waste than typical home gardens. Regardless of the composting structure and technique you use, choosing a way to process and manage plant waste is essential (see page 84).

Tool storage. You'll need a space to keep your tools protected and accessible. To maintain tools in good condition, they should be kept under cover and, ideally, out of direct sunlight. Possible locations include a garage, a tool shed, a spot under the eaves of the house, inside your propagation area, or under an elevated deck.

CHAPTER TWO

CROP SCHEDULING & RECORD KEEPING

Your goal as a production-minded gardener is not just to grow a huge amount of food but also to have it consistently available throughout the year. To accomplish this, you'll need a planting plan—a schedule specific to your garden that tells you which crops you're going to plant and when you're going to plant them. This plan, along with detailed record keeping, is what really distinguishes production gardening from ordinary backyard gardening. Crop scheduling allows you to make more efficient use of your space. In fact, you may double or triple the harvests you would normally expect from a given amount of garden space. In addition to increasing production, your planting plan will help you avoid waste from overplanting, make the best use of your soil nutrients, and minimize pest and disease problems.

Clockwise from top left: short-season 'French Breakfast' radishes; half-season 'Melissa' savoy cabbage; long-season 'Calliope' eggplant; superlong-season 'Chesnok Red' and 'German White' garlic

consider crop life span

↓ Understanding the life span and growth cycle of your crops will help you make decisions about what to plant, when to plant it, where to plant it, and how much of it to plant. Once you know how long it'll take a specific crop to grow and when it should be planted, you'll be able to determine if there's time to plant a crop before it in spring or after it in fall. You can schedule planting to make sure crops have time to mature before the end of the growing season, and to ensure you're maximizing yield by growing individual crops when soil and air temperatures are best for them. This type of understanding allows for proper planning and efficiency across an entire growing season.

When most people think of a vegetable garden, they picture annual plants—botanically defined as those that complete their life cycle in one growing season. Typically, these plants are started from seed, planted in the garden, produce a harvest, and then perish a few weeks or months later. Some annual crops like bush

A long-season crop like pumpkin can take up a garden bed for nearly 6 months.

beans die midseason, right after the edible portions are harvested. Many others, like tomatoes and squash, die as soon as temperatures drop below freezing.

Depending on where they're grown, some plants that are cultivated as annual crops can actually live for more than one season, but they're generally planted with the intention of harvesting that same season. Many herbs—such as basil, cilantro, dill, and parsley—fall into this category. Peppers and tomatoes can grow as perennials in equatorial South America, but they are almost always grown as annuals in more temperate regions such as the United States and Canada. Artichokes are perennials in the central coast of California, but must be grown as annual crops in the Northeast and Midwest, where winter temperatures are too cold for them to survive.

When you're planning what to grow, it's important to consider how long each crop will be in the ground, especially if you have limited space for growing annual plants.

Short-season crops can be grown from seed or transplant to harvest stage in a short period of time (30 to 60 days). Since they have a short growing season, you can plant these crops several times over the course of the year. Examples include arugula, cilantro, lettuce, radishes, and spinach.

Half-season crops take roughly half of a standard growing season to reach maturity (50 to 80 days). You might plant these two or three times over the course of the season: once in spring, once in early summer, and again in mid- to late summer. Examples include beets, broccoli, cabbage, cauliflower, carrots, and kale.

Long-season crops take a long time to reach maturity (70 to 120 days). You'll plant these early in the season and harvest them toward the end of the season. In many cases, you can do two plantings of long-season crops in spring, and harvest them a few weeks apart in fall. Examples include melons, peppers, tomatillos, tomatoes, and winter squash.

Superlong-season crops are the outliers to general crop categorization. They are commonly planted in fall and harvested in the summer of the following season. Examples include garlic, some types of onions, and (in some climates) fava beans.

ANNUAL CROP LIFE SPAN

CROP	DAYS TO HARVEST	CROP	DAYS TO HARVEST	CROP	DAYS TO HARVEST
Arugula	30–60, s	Collards	50–60, t	Peanut	90–150, s
Basil	60, t	Cucumbers	50–60, t	Peas, shelling	50–65, s
Beans, edible soy (edamame)	75–90, s	Dill	40–60, s	Peas, snap	50–65, s
Beans, fava (broad)	75–120, s	Eggplant	55–75, t	Peppers, hot	70–120, t
Beans, lima	60–90, s	Endive	45–60, t	Peppers, sweet	70–120, t
Beans, shell	70–85, s	Fennel, bulbing	75–85, t	Potatoes	70–120, from tubers
Beans, snap	50–60, s	Garlic	More than 120, from cloves	Radicchio	55–70, t
Beets	50–60, s	Kale	50–60, t	Radishes	30–60, s
Broccoli	60–75, t	Kohlrabi	40–80, t	Rutabagas	90–100, s
Broccoli raab	35–45, t	Leeks	70–120, t	Scallions	60–70, s
Brussels sprouts	90–120, t	Lettuce, baby mix	25–40, s	Spinach	30–40, s
Cabbage	60–90, t	Lettuce, head	30–45, t	Squash, gourds	90–130, t
Cabbage, Chinese	50–60, t	Mâche	50–65, s	Squash, pumpkins	80–120, t
Carrots	50–80, s	Melons	70–90, t	Squash, summer	45–60, t
Cauliflower	50–80, t	Mustard greens	20–25 (baby), s; 40–50 (full size), s	Squash, winter	80–120, t
Celeriac	90–110, t	Okra	50–60, t	Sweet potatoes	90–100, from slips
Celery	75–85, t	Onions, bulb	90–110, t	Tomatillos	55–75, t
Chard, Swiss	50–60, t	Pak Choi	45–60, t	Tomatoes	55–85, t
Cilantro	50–60, s	Parsley	30–60, t	Turnips	30–60, s
Corn, sweet	65–85, s	Parsnips	100–120, s	Watermelon	70–90, t

t = from transplanting, **s** = from direct seeding

Canning allows you to enjoy the fruits of your labor year-round.

planning for storage

↓   Growing storage crops is key to enjoying produce from your garden in a steady, consistent supply throughout the growing season and into winter. When you're developing a planting plan, give some thought to how and where your crops will be stored, and make selections based on your storage capacities and preferences.

Some crop varieties have much better storage lives than others. If you're planning to store a crop for a long period, make sure you choose an appropriate variety for this application. For example, the yellow onion variety 'Copra' will hold much longer than the sweet white onion variety 'Ailsa Craig'. A good seed catalog or some online research will let you know if a particular variety is well suited for storage.

When you're planning for storage, it's important to consider the work involved in putting the harvest up and the timing of each planting (see Chapter 15):

Fresh crops that must be processed for storage. These include vegetables like tomatoes, peppers, broccoli, summer squash, and snap beans. They are typically grown to be eaten fresh, but can also be canned, frozen, or dehydrated for long-term storage.

Single-planting storage crops. These include bulb onions, garlic, potatoes, dried beans, and winter squash. They are typically planted once per year, harvested, and preserved by holding them at the proper temperature and humidity (in a root cellar, for example). With enough space and proper conditions, you can eat these delicious crops well into the winter months or even into the following spring.

Keep in mind that certain varieties will last much longer in storage than others. For example, softneck garlic stores much longer than most hardneck types. Similarly, 'Yukon Gold' potatoes store for an exceptionally long time, while 'Dark Red Norland' potatoes are best eaten fresh. Plant some varieties of these crops for long-term storage and other varieties for short-term storage or fresh eating.

Succession-planting storage crops. Some crops that are planted in successions throughout the year for fresh eating are also excellent keepers in storage. Classic examples are carrots, beets, and turnips. A typical strategy for these crops is to make multiple small plantings during the growing season for fresh eating, then make a large final planting toward the end of the season

Snap beans can be pickled, canned, dehydrated, or frozen for long-term storage.

for storage. Properly stored root crops can last 6 months or longer.

Grains. It's possible to grow and store considerable amounts of carbohydrates and protein via grasses and grains like buckwheat, field corn, wheat, and rye. To produce usable yields from these sources, a large space may be necessary. Also, many of these crops become very time intensive to process and store on a small scale without specialized equipment. For these reasons, we don't discuss growing them in this book (other than as cover crops). See Resources (page 292) for more information.

Crop Scheduling & Record Keeping

31

consider your local conditions

↓ As you might imagine, crops that grow best in Louisiana may not always grow well in Michigan or Maine. Local conditions dramatically affect the success of crops and should be taken into account when selecting varieties, determining when to plant, and anticipating potential problems.

Some crops, like radicchio, are very sensitive to climatic conditions.

SCHEDULE YOUR PLANTINGS TO MATCH YOUR GROWING SEASON

The best time of the season to plant each crop varies by climate. It takes experience and local knowledge to dial in the ideal dates on a planting calendar. If you're new to an area or have never grown food in a certain climate, talk to other experienced growers or call an Extension agency. Ask for juicy secrets like how early you can plant tomatoes or what's the latest date you can sow bush beans.

If you don't have access to local knowledge for all the crops you want to plant, you can build a planting calendar based on your average first and last frost dates. An internet search of "first and last frost dates" will yield several online calculators where you can enter your location to find this information for your garden. We recommend trying a few of these online calculators and seeing how they differ, as each one may use different data to make the calculation.

PRECIPITATION & CROP YIELD

Successful growers know their local precipitation patterns and are prepared to respond, in order to get the most out of their gardens. We believe it's important for all gardeners to have a system to irrigate their crops. This becomes vital if you have significant dry spells during your growing season. For example, although the Pacific Northwest is renowned for its rainy weather, almost all precipitation comes in the fall, winter, and early spring. Gardeners must be able to irrigate continuously through the dry months of June through September. In contrast, a grower in the Midwest will often experience good rainfall throughout the growing season, but will need to irrigate during hot periods between rain events for maximum yields.

Keep in mind that all precipitation doesn't fall as gentle rain. Heavy precipitation such as thunderstorms or hail can damage crops. If you know these types of precipitation are common in your climate, be prepared to protect your crops with row cover and, if possible, time plantings to avoid damage.

Snow cover will also affect your crops. Snow can act as an insulator, protecting overwintering

To grow onions successfully, be sure to select a variety that's suitable for your latitude.

LOCAL CONDITIONS DRAMATICALLY AFFECT THE SUCCESS OF CROPS & SHOULD BE TAKEN INTO ACCOUNT WHEN SELECTING VARIETIES, DETERMINING WHEN TO PLANT & ANTICIPATING POTENTIAL PROBLEMS.

crops. It can also kill tender crops if it comes at an unanticipated time. Growers in regions that are prone to late-spring or early-fall snow events may need to rely on season-extension techniques to protect their crops.

Always be prepared for unseasonal weather events. Keep an eye on the daily forecasts and be ready to act accordingly.

LATITUDE AFFECTS GROWTH

Your position on the globe can have a surprising effect on crop growth habits. As a production-minded grower, you should learn how plants grow at different times during the season at your latitude, and plan your planting calendar accordingly.

Areas at northern latitudes will have very long days during summer, which cause crops to grow to maturity very quickly. However, they have shorter days in fall and winter, which lead to slower growth of fall plantings. Day length

dramatically affects the speed of growth of over-wintering crops.

In southern latitudes, crops may grow to maturity at a slower rate because day length is shorter than in northern regions during the growing season. However, the overall growing season will be longer, and fall and winter production is often much easier.

Some crops, like onions, require a specific amount of daylight to mature properly. Short-day onions start forming a bulb when day length reaches 12 to 13 hours. Intermediate-day (or day-neutral) onions bulb between 13 to 14 hours. Long-day onions bulb when day length reaches 14 hours or more. If you're gardening north of 40 degrees north latitude, be sure to choose long-day onions. Gardeners south of 36 degrees north latitude should be growing short-day onions, and those in between should grow intermediate-day types. Some onions are listed in catalogs as "widely adapted" and can be grown throughout the United States.

Successions of bush beans and salad greens are nestled between carrots and broccoli in this highly productive garden bed.

strategies for planting

↓   There are many ways of scheduling crops. Some growers live and die by a single method, while others use a combination of several different strategies. Consider each tactic and decide if it can be incorporated into your garden, keeping in mind that certain crops are better suited to certain strategies.

SUCCESSION PLANTING FOR A CONSISTENT SUPPLY

Succession planting—the practice of planting small quantities of a crop on a regular basis, so that they can be harvested at regular intervals—is by far the most important planting strategy for increasing garden production. Successions allow you to maintain a consistent supply of short- and half-season crops throughout the growing season.

Many new growers catch spring fever and plant an unwieldy amount of salad greens early in the season, only to be overwhelmed by their 75 heads of lettuce come June. A better approach (assuming you're growing lettuce for your own household use) would be to plant two heads of lettuce every week starting as early as weather allows and continuing until late in the season. This should allow you to harvest two heads of lettuce a week from April through October, and make a salad every night of the week.

Some short-season crops are at their peak harvest stage for a very short period, after which they quickly lose their texture and taste. Lettuce, arugula, and cilantro are good examples. These crops need to be planted every 1 or 2 weeks to ensure a steady supply of high-quality produce. Crops that can hold their quality a bit longer in the field or have a longer harvest period (beets, carrots, bush beans) might be planted every 2 to 4 weeks.

Of course, there are many situations where you might want a large quantity of a crop at one time for canning, freezing, or serving at a crawfish boil. Just make sure to size each planting appropriately for its intended end use.

PLANT IN SUCCESSION TO AVOID DISEASE

Some half- and long-season crops can be planted in successions to minimize disease problems and to maintain the longest possible harvest period. For example, many growers in the Midwest plant three or four successions of tomatoes to keep ahead of early blight. Growers in the Pacific Northwest plant two or three successions of summer squash to ensure a prolonged harvest: The later plantings come into maturity just as the early plantings succumb to powdery mildew.

SUCCESSION PLANTING WITH MULTIPLE VARIETIES

You can extend your harvest period from a single planting by using varieties that mature at different times. If you plant a 50-day, a 58-day, and a 64-day cabbage at the same time, the cabbages will be ready for harvest about a week apart from each other. Perfect! Remember, the "days to maturity" listed for a variety is usually based on ideal, controlled conditions, so the actual days from planting to harvest in your garden might be different from what is indicated on the package or catalog.

RELAY PLANTING TO USE SPACE EFFICIENTLY

Relay planting is the practice of growing two crops in the same bed, or planting a second crop in a bed where a crop is already growing, knowing that you'll harvest the first crop before the second grows too large to compete with it. If you're lucky, you might see the first crop pass a tiny baton onto the second crop right before you harvest it.

This is a great way to get crops planted earlier if you have a limited amount of space, or simply to maximize production per square foot. The only drawback is that the close spacing of crops can encourage the spread of fungal and bacterial disease. It's best to try relay planting with crops that you don't generally have disease issues with. Here are some ideas to get you started; feel free to experiment and try new ideas as you gain experience.

Snap peas. If you don't have problems with powdery mildew, try transplanting a row of snap peas along a trellis, then direct seeding a row of peas along the same trellis 1 week later. The direct-seeded peas should come into production as the transplanted ones are petering out. You can

Relay planting allows cilantro to grow under the canopy of maturing broccoli.

Salad greens and long-season transplanted crops. Place lettuce transplants in between tomato, summer or winter squash, cucumber, or brassica transplants. Plan to harvest the lettuce just before it's overwhelmed by the other, larger crop. Sowing a row of scallions, radishes, lettuce mix, arugula, or mustard greens on the outside edges of a tomato or summer squash bed will produce the same results.

Potatoes and short-season crops. Start by planting potatoes at 1-foot spacing in a bed, then direct seed arugula, spinach, or mustard greens in between the rows. You should be able to harvest the greens before the potato foliage fills out and shades the area.

Carrots and radishes. Start by direct seeding rows of carrots spaced 6 inches apart, then plant a row of radishes in between each row of carrots. Because radishes germinate and grow much more rapidly than the carrots, you should be able to harvest them about a week or two after the carrots emerge.

COMPANION PLANTING TO KEEP CROPS HEALTHY

Companion planting is the practice of growing two different crops together or adjacent to each other to save space, maximize yields, and deter pests. Make sure that you still give plants the space they need to grow well. We've seen a lot of leggy, stressed basil plants overwhelmed by sprawling tomatoes because the two crops were planted nearly on top of each other.

FOR SAVING SPACE

Corn and winter squash. Because corn grows vertically and winter squashes have a sprawling habit, the two can be grown together in the same plot to save space. Some growers believe that the scratchy leaves of the squash plants also help discourage raccoons from eating the corn. At the

also try planting pole beans along the trellis in May; they'll start climbing over the peas after the peas are done producing.

Carrots and tomatoes. Try seeding an entire bed of carrots with 6-inch spacing between the rows about 6 to 8 weeks before you transplant your tomatoes. When you're ready to plant the tomatoes, dig 8-inch-diameter holes every 18 inches down the center of the row, and transplant tomatoes into the holes. Carefully harvest the carrots with a trowel when they've sized up. If things go as planned, all the carrots will be harvested before the tomato roots start to interfere with their growth. This technique also works well with direct-seeded lettuce mix, arugula, and spring spinach.

Another kind of succession planting is simply to include many different kinds of related crops that can be used in similar ways. Onions are a good example of this. There's a wide range of onion types that prefer different growing and storage conditions but have very similar taste and nutritional benefits. In many regions, given enough space and proper planning, there is no reason you couldn't grow and eat onions every day of the year.

FIRST IN SPRING

Chives are a hardy perennial green onion, tolerant of a range of conditions. They sprout early in spring and can be harvested repeatedly throughout the season. Growing more like grass than a normal vegetable crop, a few bundles of chives planted in the garden can provide almost unlimited green onions for the duration of your growing season. These could be up and growing in spring before your cache of stored onion bulbs has been depleted.

THROUGH SUMMER

Bulbing onions are typically long-season crops that are started indoors in late winter, planted in spring, and harvested in midsummer. Depending on the variety, they will have different storage capacities. Generally, yellow types store longest. If properly cured and stored, some varieties will keep through the following spring.

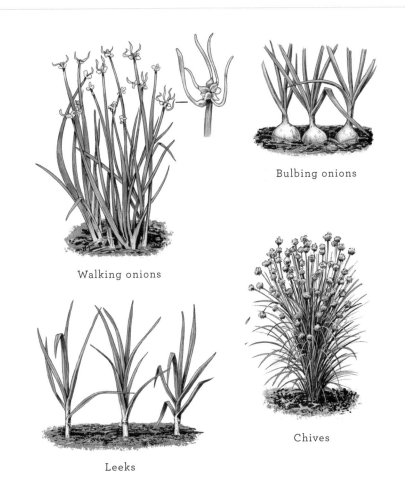

Walking onions

Bulbing onions

Leeks

Chives

FALL & WINTER

Leeks are a long-season crop; they're also very cold hardy. They're usually planted for fall harvest and used through fall and into winter. In some regions, the hardier varieties of leeks can also be heavily mulched to survive winter and be harvested in spring.

CONTINUAL HARVEST

Scallions grow relatively quickly and can be succession planted many times throughout the growing season. Since they take up little space, these green onions can fit into your planting plan in small amounts whenever needed.

Shallots are like miniature bulb onions that can be planted numerous times per year. Shallots can be planted in fall for late-winter and spring harvest or planted in spring for summer and fall harvest.

Walking onions are a perennial plant that produces an array of crops—bulbs (similar to shallots), shoots (similar to chives), and a unique "top set" of tiny airborne bulbs. Crops can be harvested spring through fall.

Scallions may help deter pests like root maggots from brassica crops.

very least, the spreading squash plants reduce the need for weeding around the corn. Make sure the soil is fertile and the plants are irrigated well, because both crops are very heavy feeders.

Corn and pole beans. Corn stalks can provide a structure for pole beans to climb. Sow two or three pole beans around each corn stalk a few weeks after the corn has emerged from the soil and begun to grow. The beans make an especially great companion for corn, since they fix nitrogen into the soil, helping the corn (a very heavy-feeding crop) get the nutrients that it needs.

FOR DETERRING PESTS

The examples listed here are not necessarily scientifically validated, but are employed by growers based on anecdotal evidence. Many growers will agree that experimenting with new crop arrangements is worthwhile even if results are not guaranteed. "Results not guaranteed" is usually the case in food production, even under the best of circumstances.

Onions and carrots. Onions (and other alliums) are believed to repel carrot rust fly. Interplant them with carrots to keep this pest away. Try spacing carrot rows 12 inches apart and planting a row of bulb onions in between the rows, or space the carrot rows 9 inches apart and plant scallions in between rows.

Scallions and brassicas. Interplanting scallions between brassica plants can help repel root maggot flies. Plant a clump of two or three scallions in between brassicas that are spaced 18 inches apart.

Cilantro and beans. Cilantro is believed to help repel aphids. Sow successions of it near crops that have aphid problems, such as brassicas and beans.

creating your planting plan

Now that you're up to speed on the techniques that production growers use to maintain a steady harvest, we can discuss how a planting plan can help you make it all happen. Your planting plan will be the backbone of your record-keeping system. Diligent record keeping will make sure the experience you gain each season can inform future choices and improve yields over time.

Making a planting plan can be daunting if you've never done it before, but once you get started, we think you'll find it to be a fun, exciting, and rewarding task. Of course, this assumes you're a total nerd like us.

To build your own planting plan, you'll need to make the following decisions in this order:

1 which crops to grow

2 how much of each crop to plant

3 when/how often to plant them

Once you've made these decisions, you'll synthesize all the data into your planting calendar. During each week of the growing season, you'll refer to your planting calendar to find out what to plant during that week. Things never go exactly as you plan in the garden, so the planting calendar has space for you to write down what you actually do (maybe you decide to plant

A planting plan will ensure you've taken the time to consider your priorities and have allocated enough space for each crop.

FROST DATES: NOT SET IN STONE

It is important to realize that the first and last frost dates indicated by the USDA or other organizations are not set in stone. In any region, frosts can happen before or after the date indicated. If you look around at various resources, you will notice that not all sources cite the same frost dates. To assess these dates, much data must be considered, and not every organization views it through the same lens.

× How deep of a frost is necessary for it to be considered a frost: 32°F (0°C), 30°F (−1°C), 28°F (−2°C)?

× Your precise location affects your frost dates. In a single city, one neighborhood may experience its last frost on March 15, while another, more protected, area sees its last frost on March 1.

× Some frost date assessments are based on different levels of risk. Is there a 10 percent chance that a frost will occur after the date noted, or a 20 percent chance?

carrots a week early, or you increase the amount of salad mix you sow each week midway through the season).

Begin by working through the Crop Amount Worksheet (page 267) and the Yields of Annual Vegetables & Herbs (page 270). These will help you calculate how many pounds of each crop you'll need to grow. Then you'll need to determine your average last and first frost dates (see the Planting Dates Worksheet, page 274) to find out your first and last planting date for each crop. If you're growing your own transplants, you can figure out when to start the seeds using the Planting Dates Worksheet.

Once your planting calendar is complete, you can decide where to plant the crops in your garden and include this information on the calendar.

HOW MUCH YOU'LL NEED

The first step is to assess your harvest goals for the garden. How much of each crop do you want to harvest each week? You can use the Crop Amount Worksheet (page 267) as a reference to help guide your thoughts. At this point, you might not know exactly how many pounds of each crop you'll need; just put down your best estimate. You might want to make a detailed plan for every crop you intend to grow, or you might create a plan for one or two important crops and wing it for the rest.

WHEN TO PLANT

Having even a general sense of the beginning and ending of the outdoor growing season will help you plan a successful garden. Knowing the length of your growing season will allow you to start transplants on time, move crops into the garden when appropriate, and protect crops from inclement weather. This is especially important if you live in a region with a relatively short frost-free period.

As you develop detailed records and years of experience, you will be the best source of local climatic data. In the meantime, consider consulting an experienced local grower who can assist you in determining target planting dates. If not, you'll have to rely on more generalized information to begin your planting calendar, such as the average first and last frost dates specified by the USDA, NOAA, or another reliable data source. Local Extension agencies will know these dates, but first and last frost dates can also be found online (see Resources, page 292).

You can enter your first and last frost dates into the Planting Dates Worksheet on page 274 and then use them to determine the earliest and latest planting dates for each crop you want to grow. As you gain experience and talk to other growers in your area, you can adjust planting dates on your planting calendar accordingly. These dates can serve as a starting point for your calculations, but should not be treated as definitive.

making your planting calendar

A planting calendar is just what it sounds like: something you can look at each week that tells you how much of each crop to plant. The calendar helps control the urge to run outside in early May and plant your entire garden with pak choi. (One of our friends actually did this, and he never really talked much about pak choi again after that. . . .) Instead, you'll check the planting calendar at the beginning of each week, then go out to the garden and calmly plant exactly what you've planned for. We recommend using one of three general formats to build your planting calendar.

A PLANTING CALENDAR HELPS CONTROL THE URGE TO RUN OUTSIDE IN EARLY MAY & PLANT YOUR ENTIRE GARDEN WITH PAK CHOI.

PRINT/ONLINE

These are great because you can quickly see what's coming up next week or next month. Online calendars are helpful because you can make one entry and set it to repeat for succession plantings. Google, Yahoo, Microsoft, or any other email provider should have a digital calendar that is easy to use and update. The drawback to these options is that there may not be a lot of extra space for taking notes or keeping track of other details.

SPREADSHEET

You can make your own calendar in the form of a spreadsheet. This allows you to create extra space to keep track of whatever information you want. You can do this by hand if you like, but a computer-generated spreadsheet program (such as Microsoft Excel or Google Sheets) is much more efficient. We generate planting calendars this way for every growing space, ranging from 200-square-foot backyard gardens to 20-acre production farms, and it works great. Use the Planting Dates Worksheet (page 274) to get started, or download the calendar template from our website (see Resources, page 292).

CLOUD-BASED RECORD KEEPING

Another option is to create your planting calendar using online software or cloud-based apps specifically designed for farms. These programs allow for very advanced and detailed record keeping, suitable for managing organic certification records and/or running a professional production farm. They typically come with an up-front and/or annual fee. Depending on your needs, they may work well, or may be unnecessarily complex.

what you'll write on the calendar

↓ At a bare minimum, you'll want to note which crop you'll be planting, how much you need to plant, and the week or date that you'll be doing the planting. This might be tracked in the number of row feet for direct-seeded crops, or in the number of plants when transplanting. You'll also need a blank space to note the date that you actually do the planting (in case it's not the same date you originally planned), and how many row feet you actually seed or actual number of transplants you set out.

Here are a few other items that are useful to note:

+ crop variety

+ exact location in the garden

+ weather conditions

+ fertilization application (quantities and dates)

PLANTING CALENDAR: A REAL-LIFE EXAMPLE

We recognize that making a detailed planting calendar sounds like a lot of work. Professional growers (like CSA managers or market farmers) might make a detailed calendar that includes every crop they're growing. If making an entire calendar sounds daunting, try it for just one or two crops that are important to you. For example, one customer we worked with was absolutely obsessed with arugula. She wasn't too concerned about scheduling the rest of her crops, but she wanted to harvest a pound of arugula per week for her entire growing season. Here's how we built the calendar:

1 The Yields of Annual Vegetables & Herbs chart (page 270) tells us 4 row feet of arugula should yield about a pound, and the Garden Planning Chart (page 284) tells us we should plant arugula once a week.

2 We determine her last and first frost dates using an online calculator, and consult the Planting Dates Worksheet (page 274) to determine the earliest and latest dates we can plant arugula: March 1 and September 15.

3 We set a weekly event in her online calendar from March 1 to September 15 that says "Sow 4 row feet of arugula, variety 'Roquette'."

No matter what kind of calendar you use and how many crops you decide to schedule, you can follow this process to make a plan:

1 Start with how much you need and how often you need it.

2 Find out when and how often you can plant it.

3 Schedule it on the calendar.

Some growers also leave space on the calendar to note their harvest data:

+ harvest date(s)

+ crop yield

+ taste characteristics

+ performance notes, including pest and disease resistance, seedling vigor, and anything else you want to remember when growing the crop in the future

If you plan to grow your own transplants, you can note when to start the seeds indoors on the calendar so the transplants are ready on time to go out to the garden. If you grow a lot of transplants, you might make two calendars: a greenhouse seeding calendar and a garden planting calendar. If you have a greenhouse or are interested in season extension, make sure to read Part 5 before making your planting plan.

EXPERIMENT WITH PLANTING DATES

There are many variables that can affect planting dates for outdoor crops, primarily your local climate and microclimate. For example, in the Pacific Northwest, it's possible to harvest broccoli side shoots from the field all winter because of the mild weather. In order to do this, the plant must have reached maturity by mid-September. This means that the summer planting date for this crop is extremely important. Use the Planting Dates Worksheet (page 274) as a starting point, but continue to finesse the calendar to find ideal dates by consulting with other local growers and by doing your own experimentation (don't forget to keep records!).

Succession planting radishes and salad greens means some spaces will appear empty as new plantings germinate next to maturing crops.

PLANTING CALENDAR WITH HARVEST TRACKING

This tracker combines the planting calendar with harvest record keeping. It's great for crops that are harvested all at one time (like carrots), but can be a little wonky for crops that are harvested multiple times over a long period (like summer squash or tomatoes). It requires that you look back in your planting calendar to note the harvest amount with the appropriate planting.

DATE PLANTED	CROP	VARIETY	ROW FEET PLANTED OR NUMBER OF PLANTS	HARVEST DATE	HARVEST AMOUNT
4/1	Carrots	'Scarlet Nantes'	Direct seed 10'	6/22	3.4 lb
4/1	Radishes	'Cherriette'	Direct seed 1'	5/1	0.4 lb
4/15	Pak choi	'Joi Choi'	4 plants	5/25	0.5 lb
4/15	Lettuce	'Coastal Star'	4 plants	5/25	1 lb
4/15	Lettuce	'Coastal Star'	4 plants	5/28	0.8 lb
4/15	Broccoli	'Fiesta'	6 plants	6/23	0.5 lb

CROP NOTES	USE AND STORAGE NOTES
Good quality and flavor. About half of the planting didn't germinate, so the yield was lower than expected. Need to keep an eye on irrigation on the next seeding.	Cut off the tops and stored in a plastic bag in the fridge.
Germinated consistently and grew quickly. Good quality, as long as they were harvested quickly.	Cut off the tops and stored in a plastic bag in the fridge. Plan to use in the next 2 weeks before they start to rot.
Cut 2 heads, each weighed about 4 ounces.	Stored in a plastic bag in the fridge.
Cut one head, the others are pretty much ready but the weather is cool. Hope to harvest one head every 2 or 3 days.	One head was enough for 2 nights of salad.
Cut the second head. A tiny bit smaller but very nice quality.	Same as above.
Cut the first head today. Five of the 6 plants are doing well. I removed one that just never took.	Used immediately.

CHAPTER THREE

GARDEN MAPPING & CROP ROTATION

Now that you've determined how much of each crop to plant and when to plant it, the next step is determining where to plant those crops in the garden. Both short-term and long-term planning are necessary for a productive garden. In any given season, you'll have to make decisions about where a crop is planted in the garden and in a particular bed.

Developing your plan ahead of time will allow you to make efficient use of your space, anticipate when space will open up for plantings later in the season, and ensure that you allot enough space for each crop, so that you can reach your harvest goals.

Understanding crop families will help you plan garden rotations more effectively. Clockwise from top left: Broccoli is a member of the Brassicaceae; tomatoes belong to the Solanaceae; and cucumbers are part of the Cucurbitaceae.

mapping the crops in your garden

↓ Drawing a crop map of the garden enables you to capture more detail about how you will use the space. This lets you keep track of where everything will go, and where you planted each crop in past seasons. If you're handy with the computer, you can make a map using free software (see Resources, page 292 for more information). Digital maps allow you to easily edit the map if you expand or make changes to your garden layout. Drawing by hand works fine, too, though. We recommend making copies of your map and using a new one each year so that you have a quick visual record of where you've planted crops each season.

Once you've finished drawing the garden map, you'll be ready to sit down and fill in where each crop from your planting plan will go. It's time to think about layout and organization.

Garden Mapping & Crop Rotation

47

TWO-YEAR CROP ROTATION PLAN

THREE-YEAR CROP ROTATION PLAN

YEAR ONE

BED 1   BED 2

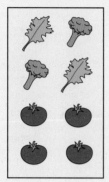   

Kale   Lettuce
Broccoli   Carrots
Tomatoes   Bush Beans

YEAR TWO

BED 1   BED 2

Lettuce   Kale
Carrots   Broccoli
Bush Beans   Tomatoes

YEAR ONE

BED 1   BED 2   BED 3

HEAVY   MISC.   HEAVY
FEEDERS   LIGHT   FEEDERS
               FEEDERS

YEAR TWO

BED 1   BED 2   BED 3

 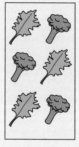 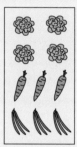

HEAVY   HEAVY   MISC.
FEEDERS   FEEDERS   LIGHT
                              FEEDERS

YEAR THREE

BED 1   BED 2   BED 3

MISC.   HEAVY   HEAVY
LIGHT   FEEDERS   FEEDERS
FEEDERS

rotate crops from year to year

↓ After deciding where to plant crops in a given season, you must plan where to plant them from year to year. Crop rotation—the practice of planting annual crops in different spaces each year—can help you create a productive long-term planting plan. Moving the planting location of each crop from year to year helps avoid pest and disease problems and uses soil nutrients more efficiently.

GROUP & ROTATE ANNUAL CROPS BY FAMILY

In order to effectively rotate crops, it's helpful to group related crops together—especially when it comes to preventing pest and disease issues and accommodating particular cultural requirements or growth habits. Some plant families include a great number of standard garden crops, so it's easy to group them together for the purposes of crop rotation.

PLANT FAMILIES THAT REQUIRE MORE FREQUENT ROTATION

BRASSICACEAE

+ **Other names:** Brassicas, cole crops, cruciferous crops, crucifers

+ **Common crops:** Broccoli, Brussels sprouts, cauliflower, cabbage, collards, kale, pak choi, radishes, and mustard greens

+ **Rotation notes:** Broccoli, Brussels sprouts, cabbage, cauliflower, and kale all have similar spacing and nutrient needs, and share many nasty soilborne pests and diseases. For this reason, we believe that these brassicas are the most important crops to rotate. We're less concerned about rotating some of the smaller brassicas like pak choi, radishes, and mustard greens.

SEPARATE SPACE FOR PERENNIAL VEGETABLES

Perennial edible crops—such as rhubarb, asparagus, artichokes, and sunchokes—live in one place for multiple years and have different management needs than annuals, so mixing the two together can complicate your crop management.

Many perennial crops tend to spread out over time. To keep perennials from spreading, it might be necessary to root prune the edges of your plantings each season, dig up and divide the base of the plant, or mow them down. Aggressive perennials like mint and horseradish should be planted in rigidly contained areas such as a bed surrounded by concrete pathways.

The good thing is that perennial crops can be incorporated into many different kinds of spaces throughout the yard. If you are interested in growing perennial edibles, take the time to fully understand their life spans, growth habits, and space requirements before placing them on the garden map (or into the garden). See Resources (page 292) for more information on caring for perennial edibles.

Apiaceae

Asteraceae

Amaranthaceae

SOLANACEAE

+ **Other names:** Nightshades, solanums

+ **Common crops:** Tomatoes, peppers, egg-plant, tomatillos, and potatoes

+ **Rotation notes:** These crops can share soil-borne diseases such as verticillium wilt and should be rotated. Though we eat their fruits and tubers, they all have mildly toxic foliage. They are all heavy feeders.

CUCURBITACEAE

+ **Other names:** Cucurbits

+ **Common crops:** Summer squash, winter squash, pumpkins, cucumbers, and gourds

+ **Rotation notes:** These crops have similar cultural requirements and share insect and disease problems. Most of their issues are not soilborne, but we still recommend planting them together and rotating them.

PLANT FAMILIES THAT REQUIRE LESS-FREQUENT ROTATION

These crops are family groups, but they have either fewer problems that can be addressed by crop rotation or very different cultural requirements and disease issues within one family.

APIACEAE

+ **Other names:** Umbelliferae, umbelli-fers, umbels

+ **Common crops:** Carrots, celery, chervil, cilantro, dill, fennel, parsley, and parsnips

+ **Rotation notes:** If you have problems with the carrot rust fly, it's good to move these crops from place to place each year to avoid overwintering larvae. Otherwise, these crops don't have a strong need for rotation. The characteristic flower structure of this family (which looks like an upside-down umbrella) is very attractive to beneficial insects. Many growers will let midseason plantings of cilantro or dill go to flower to help draw these creatures into the garden.

ASTERACEAE

+ **Other names:** Daisy family

+ **Common crops:** Lettuce, artichokes, and sunflowers

+ **Rotation notes:** Lettuce is usually clumped in with other crops in rotations, but it is important to move it each year to avoid lettuce root aphids. Some non-edible members of this family attract beneficial insects: calendula, cosmos, daisies, dahlias, zinnias, echinacea, and yarrow. Others help repel pests: chrysanthemums (a source of pyrethrin, an organic insecticide) and marigolds.

Poaceae

Amaryllidaceae

Fabaceae

AMARANTHACEAE (A.K.A. CHENOPODIACEAE)

+ **Other names:** Amaranth family, chenopods

+ **Common crops:** Beets, spinach, and chard; also includes the grains amaranth and quinoa, and the edible weeds lamb's quarters and pigweed

+ **Rotation notes:** Beets, chard, and spinach have different cultural requirements, but they share pest and disease problems (most notably downy mildew and leaf miners). In farming circles, these crops are known for having a deleterious effect on the crop that follows them in rotation.

POACEAE

+ **Other names:** Grass family

+ **Common crops:** Corn, wheat, rice, rye, barley, and millet

+ **Rotation notes:** Other than sweet corn, there are no common annual vegetables in this family. This makes small grains very useful as cover crops because they don't share diseases with most vegetable crops.

THE SPRING VS. FALL BRASSICA CONUNDRUM

One difficult choice small-scale production gardeners have to make is where to plant fall brassicas. If you're a brassica lover and you plant them in both spring and fall, you may not have enough room to plant them in two different spaces in your garden in the same year, while also maintaining a 3- or 4-year rotation.

If this sounds like you, there are two choices: Plant your fall brassicas in a different space than your spring brassicas, and only maintain a two-year rotation, or plant your fall brassicas in the same space as your spring brassicas (basically, you clear the spring planting when they've finished producing and immediately replant for fall).

The second option seems like it's breaking a lot of rules, but we actually prefer it because it makes conceptualizing and keeping track of your rotation much easier. Only go this route if you don't have problems with soilborne diseases, and be sure to apply compost and organic fertilizer between plantings.

FOUR-YEAR CROP ROTATION PLAN

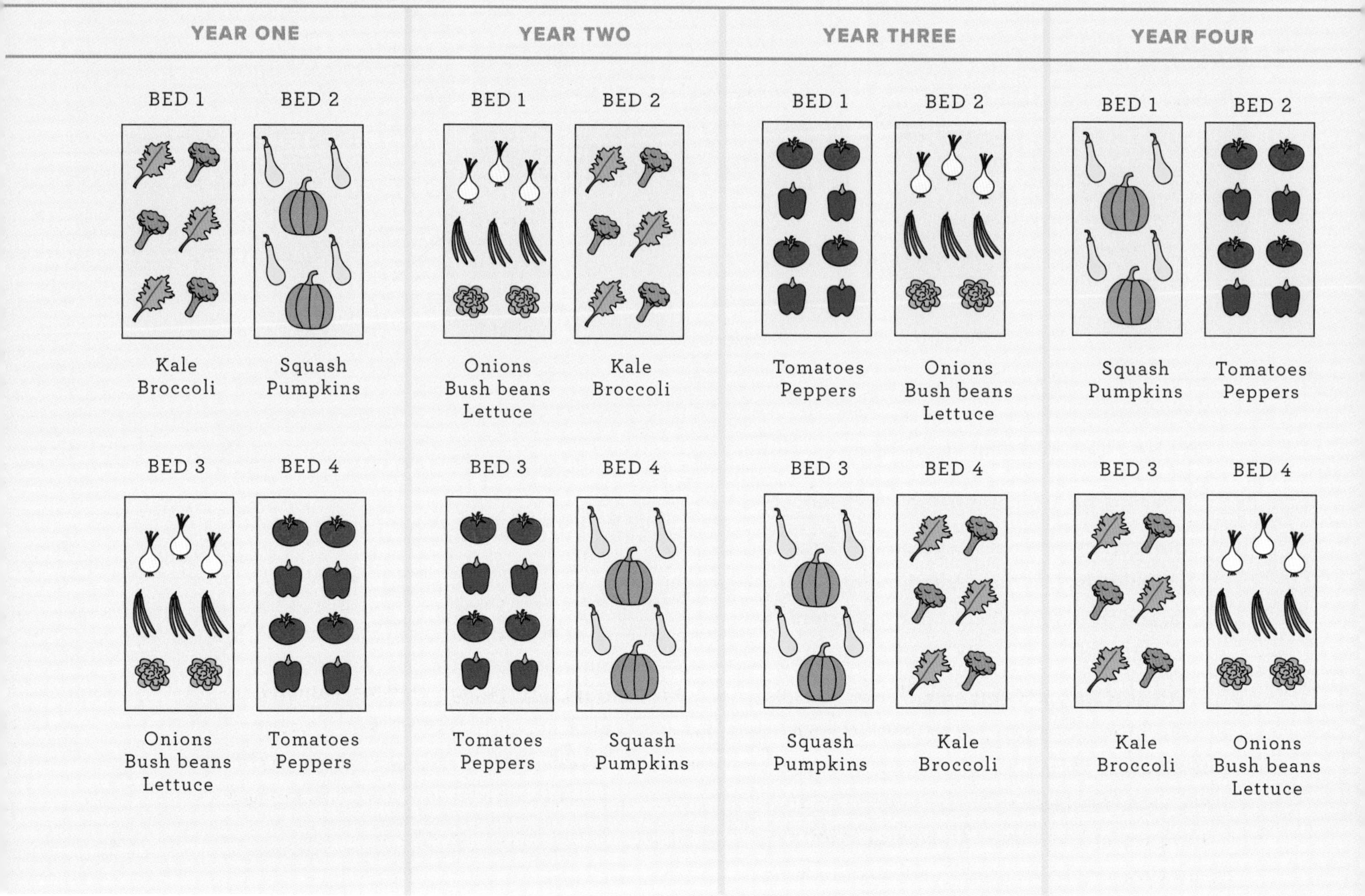

YEAR ONE	YEAR TWO	YEAR THREE	YEAR FOUR

YEAR ONE

BED 1 — Kale / Broccoli

BED 2 — Squash / Pumpkins

BED 3 — Onions / Bush beans / Lettuce

BED 4 — Tomatoes / Peppers

YEAR TWO

BED 1 — Onions / Bush beans / Lettuce

BED 2 — Kale / Broccoli

BED 3 — Tomatoes / Peppers

BED 4 — Squash / Pumpkins

YEAR THREE

BED 1 — Tomatoes / Peppers

BED 2 — Onions / Bush beans / Lettuce

BED 3 — Squash / Pumpkins

BED 4 — Kale / Broccoli

YEAR FOUR

BED 1 — Squash / Pumpkins

BED 2 — Tomatoes / Peppers

BED 3 — Kale / Broccoli

BED 4 — Onions / Bush beans / Lettuce

AMARYLLIDACEAE

+ **Other names:** Alliums (when referring to onion relatives)

+ **Common crops:** Onions, leeks, garlic, scallions, and chives

+ **Rotation notes:** Alliums are generally light feeders and don't have many disease issues, so are not major concerns in our rotation scheme. They can be susceptible to root maggots, however, so moving them every year is still a good idea. Alliums are generally considered to be beneficial to the crops that follow them in rotations.

FABACEAE
(A.K.A. LEGUMINOSAE)

+ **Other names:** Legumes

+ **Common crops:** Peas, beans, and peanuts

+ **Rotation notes:** The crops in this family are all able to fix nitrogen from the air into the soil. These are easy to include in rotations because they have few soilborne diseases, and they supply nitrogen to the following crop. Because of this, they're often used as cover crops; clover, hairy vetch, and field peas are the most common leguminous cover crops.

FERTILITY NEEDS OF CROPS

LOW		MEDIUM	HIGH	
Arugula	Fennel	Carrots	Beets	Potatoes
Basil	Kohlrabi	Celeriac	Broccoli	Squash, gourds
Beans, edible soy (edamame)	Lettuce	Celery	Brussels sprouts	Squash, pumpkins
	Mâche	Chard, Swiss	Cabbage	Squash, summer
Beans, fava (broad)	Mustard greens	Garlic	Cabbage, Chinese	Squash, winter
Beans, lima	Pak choi	Leeks	Collards	Sweet potatoes
Beans, shell	Parsley	Onions, bulb	Corn	Tomatillos
Beans, snap	Peas	Parsnips	Cucumbers	Tomatoes
Cilantro	Radishes	Radicchio	Eggplant	
Dill	Scallions	Rutabaga	Melons	
Endive	Spinach	Turnips	Peppers	

LENGTH OF ROTATION

The longer the length of the crop rotation—the number of seasons you wait before replanting a crop in the same space—the more effective it will be at helping to keep diseases at bay. For example, growing corn in one garden plot and beans in another and switching them every year constitutes a 2-year rotation of corn and beans. If you grow cabbage once every 4 years in a plot, you have a 4-year rotation. Pretty simple, right?

For the best chances of avoiding vegetable diseases, a rotation of 4 years or more is ideal, especially with brassicas. Some very skilled organic growers have up to 10-year rotation plans.

In a home garden setting, perfect crop placement and rotation can be elusive. In many cases, total garden square footage is a limiting factor. In these situations, it is usually undesirable to take portions of the garden out of rotation to allow for cover cropping and fallow seasons because this will simply reduce the number and size of your yearly plantings. Also, it's difficult to perfectly adapt a large-scale rotation plan for a limited number of beds that typically grow a diverse selection of crops. Given all that, try to understand the concept of rotation and apply it as best you can to your particular garden.

GROUPING CROPS BY FERTILITY NEEDS

Because grouping crops by family in a small production garden can be difficult (you simply may not grow enough crops in a family for them to balance out a different family in the rotation), you might choose to group crops by their fertility needs instead. For example, if you grow lots of tomatoes (nightshades) and only one winter squash (cucurbit), you might group the two together in a rotation because they're both heavy feeders. This will make it easier to apply appropriate amounts of fertilizer to the bed throughout the season. Rotating heavy feeders, light feeders, and legumes (nitrogen fixers) is a common rotation plan.

"Can you help me with my crop rotation plan?" This is one of the most common questions we hear from our friends and customers. Here's an example of one we developed for a friend. While your garden layout and favorite crops may be different, this example is helpful in understanding the concept of crop rotation and how it plays out in a real-world setting. We'll call our friend "Jason" to protect the innocent.

Eggplant and peppers are Jason's favorite crops. He's found that, in the cool northern climate he lives in, they grow best when planted right up next to the house and garage. This makes a great microclimate, because of the heat sink and southern exposure. He's never had any soilborne disease problems with these solanaceous crops, but he's dedicated to crop rotation. He rotates the two crops between the two beds each season.

One season, he plants the eggplant and peppers at the house and uses the garage bed for pole beans and onions; the next year he switches them. He also plants a few basil plants in the very southern edges of the beds in both areas because he knows they love the extra heat as well.

He splits the remainder of his favorite crops into a 3-year rotation: brassicas, cucurbits, and miscellaneous other crops (lettuce, beets, radishes, dill, bush beans). He has six beds to work with, and has labeled them 1A, 1B, 2A, 2B, 3A, and 3B to make the 3-year rotation easy to visualize.

He uses relay planting to squeeze in early carrots and salad greens around the squash and brassicas. He can't quite grow everything he wants, but he can eat fresh salads from the garden almost every week of the year. He harvests plenty of 'Czech Black' peppers, beans, broccoli, and cabbage during the growing season.

YEAR 1

1A: Brassicas

1B: Brassicas

2A: Cucurbits

2B: Cucurbits

3A: Miscellaneous (lettuce, beets, etc.)

3B: Miscellaneous (lettuce, beets, etc.)

4: Eggplant and peppers

5: Pole beans and onions

YEAR 2

1A: Miscellaneous (lettuce, beets, etc.)

1B: Miscellaneous (lettuce, beets, etc.)

2A: Brassicas

2B: Brassicas

3A: Cucurbits

3B: Cucurbits

4: Pole beans and onions

5: Eggplant and peppers

YEAR 3

1A: Cucurbits

1B: Cucurbits

2A: Miscellaneous (lettuce, beets, etc.)

2B: Miscellaneous (lettuce, beets, etc.)

3A: Brassicas

3B: Brassicas

4: Eggplant and peppers

5: Pole beans and onions

filling in your garden map

↓ Now you should be ready to start filling in your garden map and working out a rotation plan. Grab your crop checklist, a few blank copies of your garden map, and a stiff drink (whiskey, kombucha, or nettle tea, depending on your preference).

MAPPING A NEW GARDEN

If your garden is new this year, you have a blank canvas. Start by filling in locations where the long-season crops will go. Make sure tall crops won't be shading shorter crops. Next, write in where the half-season crops will go. You can write in which crops will replace them later in the season, or consider using two maps per season, one for spring and one for midsummer/fall planting. Finally, add the short-season crops. Remember, planting small amounts of these crops in succession is preferable to planting one big patch and having more at once than you can use.

If it feels weird to leave space open in the garden during spring planting, just think of the lettuce and cilantro you'll be seeding next week and the week after and the week after. Don't forget to use relay planting to fit in early plantings around many long-season crops.

Lots of detail is great, but not absolutely necessary. Many growers will make a general map of where the major crop families or specific crops will go, and work out the details on planting day—keeping a record, of course, of what was actually done.

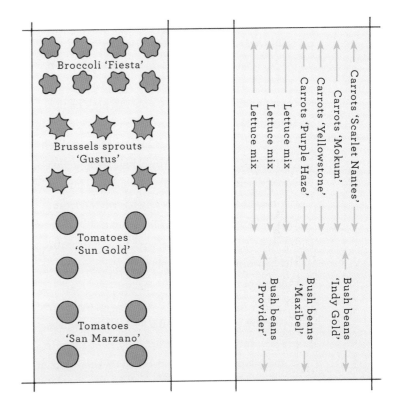

MAPPING AN EXISTING GARDEN

If you have an existing garden that has been in use for several years, the process is the same but you'll want to consider where crops were planted last year so you can get a rotation started. If you have past records, great. If not, this is the year to start keeping them.

As you're working through the process, a long-term rotation plan might begin to make more sense. If so, you can fill out future garden maps now, noting where crop families will be placed for the next several seasons. If not, that's okay. After a few seasons of practice, a long-term plan might become apparent, or you can continue to rotate on a year-by-year basis.

keeping additional records

↓ Keeping accurate records is a powerful way to expand your knowledge and refine your growing practices so that you can improve your yields from one year to the next. If you keep good records, you can look back through the years and see which soil amendments improved yield and which had no effect, which varieties of beans performed best, which pest control practices actually worked, and how many pounds of spinach you really harvested from that 8-foot row. You can also keep track of time between planting and harvest, and determine the best time of year to plant crops in your specific climate.

Your style of record keeping can take any form; you should develop a system that works for you and that you can actually keep up with. Some people write their records on their garden maps. Some people keep their records right on their planting calendar. Some people write everything down in a regular old notebook. Some build spreadsheets on a computer. And some people use cloud-based record-keeping applications.

Here are the basic pieces of garden and crop information we recommend keeping in your records:

+ Date of harvest for each planting

+ Size of the planting (how many row feet or the number of transplants)

+ Location of planting: the specific field, bed, or part of a bed that was planted

+ Yield (how many pounds you harvested)

+ Flavor (was it worse, as good, or better than expected?)

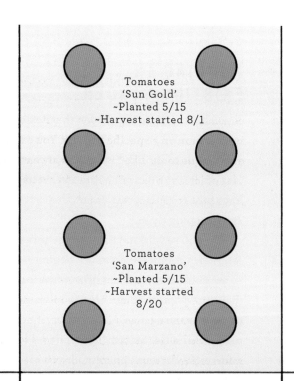

Tomatoes
'Sun Gold'
~Planted 5/15
~Harvest started 8/1

Tomatoes
'San Marzano'
~Planted 5/15
~Harvest started 8/20

The level of detail you provide on the garden map can vary. You might draw a garden map to scale and note exactly where every plant will go. Such a drawing might look like this.

- Relative performance of different varieties. If certain varieties do not produce well in your garden, you'll want to note this and remove them from your planting plan the following season. If a variety has standout yields, superior pest resistance, or excellent flavor, you'll want to grow more of it next season.

- The soil amendments and organic fertilizers you used, and in what quantities

- Any pest or disease problems and which control methods you used (and whether they were effective)

- General observations throughout the season. This might include information like, "'Red Russian' kale definitely had the best flavor of all the varieties I tried this year," or "The beans I seeded on August 5th didn't have time to mature; I'm going to put my last planting in a little earlier next year."

We also recommend that you keep track of a few other items:

- Seed suppliers, and varieties and quantities you order. This will help you put together next season's seed order. (See Chapter 7 for more on seeds.)

- Suppliers for tools, amendments, and organic pest and disease control materials. If you're interested in tracking costs, keep receipts for these supplies.

- If you use municipal water for irrigation, you might keep your water bills so you can track how much water you use in the garden each season. Similarly, if you use electricity for grow lights or greenhouse heat, keep track of the correlated utility bills.

VISUAL RECORD KEEPING

Many growers like to keep track of plantings directly on their garden map. This is a good way for visual thinkers to see what's happened in a particular bed or field over the years. If you want to keep track of a lot of information, this method

A WEATHER-RESISTANT NOTEBOOK

If you plan to keep any of your records on paper (instead of on a computer), we recommend one that is designed to withstand moisture and sun exposure. These books typically have thicker paper and a water-wicking cover and will comfortably put you in the small cadre of nerds that also includes birders and herpetologists.

may be limiting because you won't have much space on your map to write things down. However, it may provide an opportunity to draw pictures of dinosaurs or yetis lurking in the margins.

ACTIVITY LOGS

Some growers like to make a separate log for each different activity. For example, you might have a harvest log, a soil amendment log, a pest and disease control log, and a planting calendar. This means you'll have more logs to maintain, but it is useful for large-scale growers with a lot to keep track of. This is the style of record keeping that many inspectors prefer if you are planning to be certified organic. It allows you to quickly gather information on a specific activity. You can easily answer questions like "What was my earliest harvest of lettuce this spring?" or "What did I spray for powdery mildew last year?"

ALL-IN-ONE LOG

Some small-scale gardeners keep track of everything in one general log. This keeps things simple because you have fewer logs to keep track of, but it's a little harder to look back and find specific information because individual activities aren't differentiated.

BED-SPECIFIC LOG

If you want to use your map to keep records but would like more space to write things down, dedicate a page in your notebook or tab in your spreadsheet for each bed or field in your garden. A bed-specific log will allow you to easily look back at past seasons and see what you planted in each bed and when it was harvested. It'll provide a simple reference to note days to maturity and other information that varies by location.

BED NUMBER: 1A

DATE	ACTIVITY
5/1	Transplant 'Sun Gold' cherry tomatoes, cover with row cover.
6/3	Remove row cover, prune, side-dress w/5:5:5 fertilizer.
7/1	Foliar feed w/liquid kelp.
7/16	Foliar feed w/liquid kelp.
7/24	First harvest.
9/1	Signs of late blight! Pull plants, clean all dead foliage, burn.
9/7	Direct seed 'Regiment' spinach.

PEST AND DISEASE CONTROL LOG

DATE	PEST/DISEASE OBSERVED	LOCATION	CROP AFFECTED	MANAGEMENT APPLIED	RESULT
7/13	Imported cabbageworms	East field	Broccoli	Spray spinosad.	7/22: damage appears to have stopped.
9/16	Slugs	Northeast plot	Spinach	Spread diatomaceous earth.	9/18: rain washed off DE, repeat application.

SEATTLE URBAN FARM COMPANY GARDEN LOG

DATE	HOURS SPENT	PLANTING NOTES	FERTILIZATION NOTES	PEST/DISEASE/ PLANT HEALTH ISSUES	OTHER NOTES	TO-DO LIST
3/15	1.5	Planted potatoes in bed 1. Direct seeded snap peas in bed 2.	Added compost to beds 1, 2, and 3.	None	Checked pH: bed 1: 6.5 bed 2: 6.7 bed 3: 6.4	Buy more fertilizer.
4/2	2	Planted kale, broccoli, cabbage, and cauliflower in bed 3. Planted lettuce mix next to pea trellis in bed 2.	Spread 3:3:3 before planting.	None	Turned on drip system at 3x/week, 30 minutes per run.	Peas haven't germinated yet; keep an eye on them.
4/12	1	Planted more lettuce mix, cilantro, carrots, turnips, and radishes in bed 3.	Spread 3:3:3 before planting.	Brassicas appear to have slug damage. Spread iron phosphate slug bait in beds 2 and 3.	Peas finally germinated.	Order more lettuce mix seed; running low.
4/19	30 minutes		Sidedressed potatoes with 3:3:3.	None	Weeded all beds. Tied peas to trellis.	
5/1	2	Planted tomatoes under row cover in bed 3. Direct seeded more lettuce mix in between tomato rows. Planted broccoli in bed 2.	Spread 3:3:3 before planting. Side-dressed brassicas in bed 2 with 3:3:3.	Imported cabbageworms in brassicas. Sprayed spinosad.	Hilled potatoes.	

PART

2

BUILD HEALTHY SOIL

↓

MAXIMIZE HARVESTS BY PROVIDING SOIL WITH HIGH NUTRIENT LEVELS & A DIVERSE ECOSYSTEM.

Now that we've discussed how planning and record keeping can increase yields, let's consider actions you can take out in the garden to do the same. This includes managing the health of your soil, solving pest and disease problems, and suppressing weeds.

CHAPTER FOUR

PREPPING SOIL FOR PRODUCTION

If you don't have good soil, you're going to have a tough time growing productive crops. Taking the time to create beds full of rich, fertile soil will put you solidly on the path to high yields. Whether you're expanding your growing area with new beds, or improving the beds you already have, it's important to understand the process of creating healthy, productive garden soil.

building new garden beds

If you're building a new garden or adding new beds to expand an existing garden, it's critical to put in the initial effort to create nutrient-dense soil with good structure—this is the foundation for maximizing your yield. If you already have well-established garden beds, this section may not apply to you.

Before you can start setting up new beds, you'll first need to eliminate the competition. There are two basic approaches—you can either clear the site completely, or you can incorporate existing materials into the soil. Each technique has its pros and cons.

CLEARING THE SITE

If your garden site is full of undesirable plants, such as invasive weeds or large shrubs, you'll need to clear the entire site as the first step in garden building. Removing large plants, random debris, grass sod, rocks, invasive weeds, and ground covers will set you up with a clean slate. This clearing will make it easy to work the soil and determine how much soil amending is necessary. Site clearing also provides a great opportunity to regrade the space as needed to prevent erosion and make level beds and pathways.

LOOSEN THE SOIL

After clearing the site, loosen the soil with a sod fork, shovel, or broadfork. It's not necessary to turn the soil over; simply lift the soil and shake it a bit to break up clods. This will typically reveal more roots and rocks that can be removed from the soil. After thoroughly loosening the soil, decide where your garden beds and paths will be and stake them out.

Structured raised beds offer deep, loose soil and good drainage.

SOURCING QUALITY BULK MATERIALS

As you expand your production, you're likely to need larger quantities of amendments. Anticipating the materials you'll need during the course of a season will help make sure you have time for sourcing, pickup, or delivery of any given material before it's required. The compost and amendments you use may depend primarily on what is available to you. We know an organic farmer who applies only cow manure to his beds, because he has an unlimited supply of this free resource. We know another who fertilizes exclusively with pelletized chicken manure because he finds this the easiest and cheapest option for his situation. If you're lucky enough to have access to free, high-quality materials, choose those. Otherwise, look for professionally made organic compost and fertilizer made from local ingredients when possible.

Let's take a look at some of the most common soil materials and how to approach acquiring them.

SOIL

Do your research and try to locate a place that has a good reputation and understands the needs of an organic vegetable gardener—you'd be surprised how many don't. If you're buying new soil for your garden beds, ask for a two-way soil mix made of compost and sand, or a three-way soil mix made from compost, sand, and topsoil. Soil purveyors will often carry other useful products like compost, gravel, and mulch.

MANURE

With some searching you may find sources of manure that are available for free. Make sure the manure is fully (and appropriately) composted before adding it to your garden beds. Fresh, uncomposted manure may be easier to find, but can add pathogens and weed seeds to your garden. If you do source uncomposted manure, take the time to compost it yourself before adding it to your garden.

SOIL AMENDMENTS & FERTILIZERS

You may need a variety of soil amendments for your garden (see page 78). If you have the capacity to purchase and store these materials in larger quantities, you can save quite a lot of money. Often a 50-pound bag of a basic soil amendment will cost little more than a 5-pound bag from the garden center. Look for wholesale distributors or local businesses that are willing or able to deal with larger volumes, and consider splitting orders with other local gardeners.

AMEND THE SOIL

At most sites, the existing soil will need additional compost or garden soil to create a desirable structure, nutrient base, and depth for vegetable crop production (see Chapter 5 for more information).

Start by spreading a 4- to 6-inch layer of high-quality compost or vegetable garden soil mix onto the annual beds. Mix this into the existing soil with a shovel or garden fork to create an entirely new soil structure. After adding the new compost/soil, take a soil sample for testing or test the pH yourself to determine what other amendments to add (see page 69).

The perennial edibles that benefit most from soil amendment are asparagus, artichokes, rhubarb, blueberries, raspberries, and strawberries. Less-demanding perennials, such as fruit trees, most herbs, and sunchokes, can be grown with little soil amending. Some growers suggest that planting these perennials in native soils without amendment allows them to adapt better to local conditions and thrive. In any case, you'll still need to loosen the soil and remove roots and rocks. This technique won't work if you have extremely sandy, clayey, or otherwise low-quality soil. If your existing soil is very poor, prepare perennial beds in the same way you prepare your annual beds, as detailed above.

INCORPORATING THE MATERIALS ON-SITE

If your new garden site is currently a grass lawn, small weeds, or a nonaggressive ground cover, you might simply till the existing plants into the soil or smother them until they've decomposed. Incorporating these plants will help increase your soil's nutrient levels and add organic matter, and will eliminate the need to haul materials off-site.

When incorporating on-site materials, start the process at least 4 weeks before planting. If you're able, get started 8 or more weeks ahead of planting. This gives the plants you're tilling in or smothering time to decompose before crops go in the ground. Planting in the garden too

early will be awkward and inefficient because of large clumps in the soil. It will also result in poor productivity because soil nutrients will be unavailable due to the decomposing residue. Weeds or ground covers with woody stems should be removed; they will take far too long to decompose.

TILLING

The right tool for tilling your site will depend on the scale of the project and your gusto for manual labor. Annual weeds are relatively easy to break up; grass sod is extremely difficult. A sod fork or shovel works for very small spaces. For turning in large areas of grass sod, we strongly recommend using a hydraulic rear-tine rototiller. It might sound intense, but it's easy to use and can be rented from most equipment rental stores. To make your job easier, be sure to mow the grass/weeds as short as possible before turning them in. If the soil is dry, water it with a sprinkler before trying to work it. Dry soil is incredibly difficult

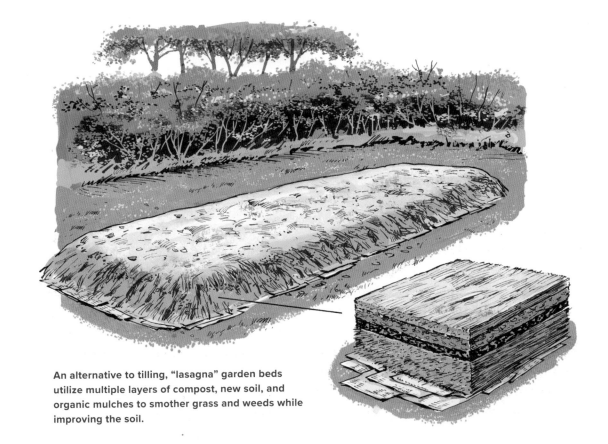

An alternative to tilling, "lasagna" garden beds utilize multiple layers of compost, new soil, and organic mulches to smother grass and weeds while improving the soil.

to work. Don't overdo it, as wet soil will clump up and create persistent clods if you work it. Aim for a handful of soil to feel like a damp sponge before you till or fork.

Chances are, you'll need to work the soil at least two or three times, especially when trying to break down heavy grass sod. Till everything in, wait 2 to 3 weeks, and then till again and wait a few more weeks. If the plants have broken down well and there's no residue, proceed with building your garden. If residue persists, wait longer or turn again. If the soil is dry during this process, water it to promote decomposition.

SMOTHERING EXISTING PLANTS

Another method of incorporation is commonly called "layer" or "lasagna" gardening. The idea is that, with the addition of enough organic matter—typically layers of cardboard, compost, and soil—you can smother the existing plants and create new beds directly on top of them. If the new beds are topped with a thick enough layer of soil and compost (6 to 12 inches), you can plant right away as the under layers continue to decompose. You can add additional soil, sand,

or compost to improve the structure of the soil, and you should test the pH/professionally test the soil to determine if any other amendments are needed. See Resources (page 292) for more information.

NO-TILL TARPING

You can also smother existing plants using a heavy plastic tarp. Simply place a heavy opaque tarp (5- to 6-mil silage tarps or old pond liner works great) over the area you want to make into a new garden bed. Weight the tarp with sandbags, rocks, or other heavy items, and leave in place until the existing plants are killed and have broken down. The benefit of this method is that it maintains great soil moisture and worm activity is increased due to the darkness, so when you remove the tarp the soil is in great condition and the old plants are fully decomposed with a minimal amount of labor. The downside is the time it takes—anywhere from a few months to over a year for well-established perennial grasses. Also, leaving a tarp in place in your garden may encourage rodent activity. To learn more about using tarps to weed existing beds, see page 99.

NO-TILL TARPING IS AN EFFICIENT WAY TO BREAK DOWN CROP RESIDUE IN BEDS AFTER HARVEST.

Tarps help reduce competition from weeds and speed up decomposition of crop residue.

MAXIMIZE EFFICIENCY: MULCH YOUR PATHWAYS

Whether you're starting new beds or improving old ones, we recommend you mulch your pathways to reduce erosion and time spent weeding (not to mention keeping your shoes and pants cleaner). There are any number of mulch materials you could use, depending on your preferences. See the chart below for a few ideas to get you started.

Instead of using nonliving materials, some growers plant their pathways with a cover crop, as a "living mulch." If you go this route, be sure to mow the cover crop before it sets seed or it will become a weed problem unto itself. Avoid using perennial grasses for living mulch unless you have timber-framed or other tall, enclosed raised beds. Otherwise, the grass will spread by its rhizomes into your garden beds. See page 89 for more information on cover crops.

MATERIAL	RECOMMENDED?	NOTES
Bark mulch	Yes	Must be replenished each year to suppress weeds effectively. In addition to smothering weeds, bark mulch often contains hydroxylated aromatic compounds, naturally occurring chemicals that have an allelopathic effect on weeds (discourage their growth).
Wood chips	Yes	Must be replenished each year to suppress weeds effectively. In addition to smothering weeds, some wood chips (such as cedar) contain natural chemicals that help discourage weed growth.
Straw	Yes	Must be replenished each year to suppress weeds effectively. Consider using with an underlayer of newspaper or cardboard.
Gravel	With caution	Long-lasting, but weeds grow through it quickly and are difficult to remove. Best used when applied on top of an underlayer of newspaper, cardboard, or heavy landscape fabric.
Newspaper	With caution	Breaks down quickly, presents a trip hazard, and is unsightly. Best used as an underlayer, topped with bark mulch, wood chips, or straw.
Cardboard	With caution	Effective at blocking weeds, but is annoying to walk on and can acidify soil as it breaks down. Not appropriate for certified organic gardens. Best used as an underlayer, topped with bark mulch, wood chips, or straw.
Heavy landscape fabric	Yes	Look for a "20 year" rating. Can be rolled up in winter for better longevity. Can be difficult to weed around if weeds get out of control.
Light landscape fabric	No	Breaks down too quickly, wasting time, money, and resources. Weed roots can grow into the fabric, making it difficult to remove.
Plastic	No	Tears apart, creates garbage, and is unsightly.

improving existing beds

↓ If you have an existing garden that has not been managed with an eye toward ongoing soil management, you likely have depleted some (or most) of the essential nutrients. You many also have an overly acidic or basic pH. Plants absorb significant quantities of nutrients from the soil each year, and unless you are actively replacing these nutrients, plant vigor and yield will decrease over time. Fortunately, it's never too late to begin rebuilding garden soil, and significant improvements can be made relatively quickly.

SOIL FERTILITY

Many gardeners maintain reasonably good soil fertility simply by adding compost, using organic fertilizers, and ensuring that their soil pH is in a reasonable range (between 6 and 7

for most crops). However, if your goal is to maximize your yields using organic methods, a professional soil test will help make sure your nutrient and pH levels are optimal, and will help you identify any trouble spots before they become a problem.

If compost is the only soil amendment you use, your crop productivity and nutrient density is heavily dependent on the quality of the compost. In our experience, the quality of compost varies widely, regardless of whether it's homemade or commercially made. Without testing, there's no way to know if your soil and compost contain the full complement of minerals and nutrients necessary to grow quality crops. We have seen many gardeners and farmers get good results for a few years simply by adding compost and organic fertilizer, only to be plagued by poor

PLANTS ABSORB SIGNIFICANT QUANTITIES OF NUTRIENTS FROM THE SOIL EACH YEAR; UNLESS YOU ARE ACTIVELY REPLACING THESE NUTRIENTS, PLANT VIGOR & YIELD WILL DECREASE OVER TIME.

Soil testing can help you create an environment where even finicky crops like cauliflower can thrive.

production as time goes on and specific nutrient deficiencies become more pronounced.

In addition, different soils can have incredibly different latent nutrient bases, and the lack of a single micro- or macronutrient can dramatically affect the productivity of your garden. Plants do show visible signs of nutrient stress, but these deficiencies can be difficult to correctly identify, especially when you are dealing with micronutrient issues. Identifying your specific soil nutrient levels is a critical first step toward efficiently remineralizing your soil to improve crop yield and nutrient density. And for you single gardeners out there, discussing your soil's cation exchange capacity is a great way to pick up a date at the next potluck dinner.

START WITH A SOIL TEST

Fall is a great time to test your soil because it gives you time to make changes before spring planting. Early spring also works well.

If you're building a new garden, we recommend testing your soil after you've added new soil or compost. Adding large amounts of soil or compost will change your pH and nutrient levels. If you test before you've added soil or compost, your test results will not reflect the changes you've made and will guide you to add more organic fertilizer and other mineral amendments than necessary.

If you have added organic fertilizer or rock powders to your soil, avoid testing for 6 to 8 weeks because nutrient levels will be temporarily altered.

TAKE A SAMPLE

First, select a lab to send your sample to (see Resources, page 292, for more information). Once you've chosen a lab, read their guidelines for collecting your sample.

To collect your sample, use a trowel or spade shovel to dig eight to ten small holes about 6 inches deep in different spots. Put a slice of

(see Resources, page 292, for more information)

TEST REGULARLY

For most situations, testing the soil once every 3 years is plenty. If you're really interested in learning how your amendments are affecting the soil each year, try testing once a year for 3 years. You can then test less frequently in the future if your soil is changing the way you want it to.

Over time, you might see that your soil test shows that your garden has high levels of soluble salts. These are usually caused by excessive applications of organic fertilizer. It's a good reminder to be sure you apply organic fertilizer as recommended. If your soil tests high for soluble salts, you can leach them out by running 2 to 4 inches of water through it. You can do this by running your irrigation system with drip tape for 12 to 15 hours. If you use a sprinkler, run it until water pools on the surface of the soil, turn it off, let the water soak in, and repeat four to six times. Note that this can deplete other, desirable nutrients as well.

soil from the edge of each of these holes into a plastic bucket. The slices should represent the entire depth of the hole—they should include soil from 1 inch down, 2 inches down, and all the way down to 6 inches. Remove any plant fragments or large chunks of organic matter. Mix well, remove 1 cup's worth, and bag, label, and send to the lab, per their specifications. When you send it in, make sure you specify that you'd like recommendations for organic management and that you are growing vegetable crops. This will help make sure the information is tailored to your needs.

For most small production gardens without major issues, one test for the whole plot will yield plenty of information. If you have several garden plots that you've been managing differently or have different soil types, you will want to test each plot individually. Similarly, if there's one part of your garden that's having issues, such as plant discoloration or stunting, sample it separately from the rest of the garden.

how to interpret soil test results

↓ Ah, here's the fun part! Featured on the facing page are the results of an actual soil test we took for a small production farm (½ acre) with three different planting areas (labeled North Main, South Main, and East).

Always remember to take soil test results with a grain of salt; sampling errors can yield strange results, and labs can make mistakes. If a soil test shows results that don't correlate to the behavior of your plants, consider doing a second test or trying a different lab. For example, if a test result indicates that your soil pH is extremely low but your plants are growing fine and a test from last year showed your pH to be in a normal range, consider sending in a new sample to be tested.

1. Recommendations. This is where the lab tells you what nutrients to add to the soil so crops thrive.

2. N:P:K. Nitrogen (N) is responsible for early vegetative growth in plants. Phosphorus (P) is crucial for photosynthesis, early root growth, and flowering. Potassium (K) helps with photosynthesis, fruit formation, and disease resistance. The lab has provided a convenient scale so it is easy to tell if your soil is deficient in any of these key nutrients. The Recommendations page (1) gives us amounts of these nutrients to add in pounds per acre.

An important item to note is that available nitrogen can vary on a month-to-month and week-to-week basis depending on soil temperatures and precipitation, so what we see in the test may not be what's actually happening in the soil today. Nitrogen testing is an add-on for this lab, so we decided not to test for it.

3. Calcium and magnesium. An important item to note is that calcium and magnesium should exist together at a ratio of around 5:1 to 7:1 (Ca:Mg). If the ratio gets out of whack, both nutrients can become difficult for plants to absorb. The lab should know this, and it should reflect in their recommendations, but it's good to be able to check on your own. Calcium and magnesium levels look good for all the plots in this test.

4. Sulfur, sodium, manganese, iron, copper, boron, and chloride. These are micronutrients, and are important for plant growth but only in very small amounts. Typically you won't need to be concerned about these nutrients unless there is a noticeable deficiency in the test results.

One exception is that brassica crops are susceptible to sulfur and boron deficiencies, so keep an eye on these levels if you grow a lot of brassicas. In this test, our sulfur, manganese, zinc, and boron levels look low, so we'll want to boost them (see How We Took Action, page 72).

5. Organic matter. In a large-scale agricultural setting, 5% organic matter is a great number, and 3 to 4% is not bad. Expect a higher percentage of organic matter in clay soils, and lower percentage in sandy soils. For those of us growing in intensively managed raised beds or in compost-based potting soil, you might see numbers in the 20 to 30% range. As long as you're at 5% or higher, you don't need to be concerned about this specific number; organic matter is always breaking down and needs to be replenished regularly using compost or cover cropping techniques. Organic matter levels look great in this test.

A&L Eastern Laboratories

7621 Whitepine Road Richmond, Virginia 23237 (804) 743-9401 Fax (804) 271-6446

Send To: SEATTLE URBAN FARM COMPANY

Grower:
SEATTLE URBAN FARM COMPAI

Submitted By: BRAD HALM
Farm ID:

Date Received: 01/24/2014
Date Of Report: 01/27/2014

SOIL FERTILITY RECOMMENDATIONS

Sample ID Field ID	Intended Crop	Yield Goal	Lime Tons/A	Nitrogen N lb/A	Phosphate P₂O₅ lb/A	Potash K₂O lb/A	Magnesium Mg lb/A	Sulfur S lb/A	Zinc Zn lb/A	Manganese Mn lb/A	Iron Fe lb/A	Copper Cu lb/A	Boron B lb/A
NORTH MAIN	Garden, Market	0	1.5	60	200	200	0	27	3.0	3	0	0	2.5
SOUTH MAIN	Garden, Market	0	1.0	60	50	250	0	31	3.0	3	0	0	3.0
EAST	Garden, Market	0	0.0	60	40	118	0	27	0.7	0	0	0	3.0

Comments:

Sample(s) : SOUTH MAIN,EAST Crop: Garden, Market

On market garden apply an additional 40-100# of N per acre sidedress using higher rates for green, leafy vegetables, peppers, tomatoes, sweet corn, etc. and lower rates for peas, beans, melons,cucumbers,carrots, root crops, etc. On tomatoes do not apply additonal N until fruit set are the size of a golf ball.

"The recommendations are based on research data and experience, but NO GUARANTEE or WARRANTY expressed or implied, concerning crop performance is made."

Our reports and letters are for the exclusive and confidential use of our clients,, and may not be reproduced in whole or part, nor may any reference be made to the work,the results, or the company in any advertising, news release, or other public anouncements without obtaining our prior written authorization. Copy right 1977.

Pauric McGroary

Pauric McGroary

A&L Eastern Laboratories

7621 Whitepine Road Richmond, Virginia 23237 (804) 743-9401 Fax (804) 271-6446

Send To: SEATTLE URBAN FARM COMPANY

Grower:
SEATTLE URBAN FARM COMPAI

Submitted By: BRAD HALM
Farm ID:

SOIL ANALYSIS REPORT

Analytical Method(s):
Mehlich 3

Date Received: 01/24/2014   Date Of Analysis: 01/27/2014   Date Of Report: 01/27/2014

Sample ID Field ID	Lab Number	Organic Matter %	Organic Matter Rate	Organic Matter ENR lbs/A	Phosphorus Mehlich 3 ppm	Phosphorus Mehlich 3 Rate	Phosphorus Reserve ppm	Phosphorus Reserve Rate	Potassium K ppm	Potassium K Rate	Magnesium Mg ppm	Magnesium Mg Rate	Calcium Ca ppm	Calcium Ca Rate	Sodium Na ppm	Sodium Na Rate	pH Soil pH	pH Buffer Index	Acidity H meq/100g	C.E.C meq/100g
NORTH MAIN	22540	6.3	H	150	19	L			115	M	190	M	1295	M	43	VL	5.6	6.67	2.6	11.1
SOUTH MAIN	22541	6.7	H	150	56	H			94	M	180	H	1148	M	49	VL	5.8	6.75	1.8	9.5
EAST	22542	6.0	H	150	130	VH			149	H	219	M	1703	M	63	VL	6.3		1.3	12.3

Sample ID Field ID	Percent Base Saturation K %	Percent Base Saturation Mg %	Percent Base Saturation Ca %	Percent Base Saturation Na %	Percent Base Saturation H %	Nitrate NO₃N ppm	Nitrate Rate	Sulfur S ppm	Sulfur Rate	Zinc Zn ppm	Zinc Rate	Manganese Mn ppm	Manganese Rate	Iron Fe ppm	Iron Rate	Copper Cu ppm	Copper Rate	Boron B ppm	Boron Rate	Soluble Salts SS ms/cm	Soluble Salts Rate	Chloride Cl ppm	Chloride Rate	Aluminum Al ppm
NORTH MAIN	3.0	14.0	58.0	2.0	23.0			17	M	2.0	L	7	L	400	VH	3.4	VH	0.6	M					
SOUTH MAIN	3.0	16.0	60.0	2.0	19.0			14	L	2.0	L	9	L	411	VH	3.0	H	0.1	VL					
EAST	3.0	15.0	69.0	2.0	11.0			17	M	6.7	H	24	H	376	VH	2.2	H	0.2	VL					

Values on this report represent the plant available nutrients in the soil. Rating after each value: VL (Very Low), L (Low), M (Medium), H (High), VH (Very High). ENR - Estimated Nitrogen Release. C.E.C. - Cation Exchange Capacity.

Explanation of symbols: % (percent), ppm (parts per million), lbs/A (pounds per acre), ms/cm (milli-mhos per centimeter), meq/100g (milli-equivalent per 100 grams). Conversions: ppm x 2 = lbs/A, Soluble Salts ms/cm x 640 = ppm.

This report applies to sample(s) tested. Samples are retained a maximum of thirty days after testing.

Analysis prepared by: A&L Eastern Laboratories, Inc.

by: *Pauric McGroary*

Pauric McGroary

6. Soil pH and buffer pH. Vegetables like a narrow pH range; 6.3 to 6.8 is ideal. Buffer pH is an artificial number the lab has created to calculate how much lime or sulfur you need to add to effectively change the pH. Generally speaking, you can use the buffer pH to calculate how much lime or sulfur to add. In this test, the actual pH is a little low, but the buffer pH is in a good range. The lab suggests adding lime in the recommendations, so we'll go with what they say.

7. CEC (cation exchange capacity). This is a measure of the soil's ability to both store nutrients (specifically cation nutrients) and make them available to our crops. The higher the number, the higher the storage capacity of the soil, but the more difficult it is for the soil to make the nutrients available. A high number also means you'll need to add more of an amendment to make a change. The lab knows this, so it should be reflected in the recommendations.

Generally, as you add compost to the soil, your CEC will rise. Well-composted soils are often in the 15 to 25 range. Clay soils also have a very high CEC (25 to 50), but will still benefit from the addition of compost. You don't need to worry too much about this number unless it's below 10. Lower numbers mean you have sandy soil that would benefit from the addition of extra compost. We added compost to the South Main field, since it had a reading of 9.5.

8. Percent base, nutrient, or cation saturation. This number tells us how "filled" our particular soil is with certain nutrients. Using these numbers to make recommendations is beyond the scope of this book. If you want to learn more, check out Resources (page 292).

OTHER ITEMS YOU MIGHT SEE ON A SOIL TEST

Soluble salts. Overapplication of organic or synthetic fertilizer can cause salt toxicity in some plants. Look for a range of 0.08 to 0.5 dS/m. Over 0.6 dS/m can cause problems such as poor germination, browning leaf margins, yellowing, and stunted growth.

Lead, aluminum, arsenic, heavy metals. These are elements that can be toxic to humans or plants if levels are too high. Lead and arsenic can be especially important to test for if you're farming in an urban area, because there are many potential sources of these contaminants. (See Dealing with Lead & Arsenic Contamination, page 74.)

HOW WE TOOK ACTION

N:P:K. Because we always add balanced organic fertilizer to our soil just prior to planting and side-dress heavy-feeding crops (see page 141), we don't usually worry too much about these numbers. We could spread rock phosphate in the North Main field because the phosphorus tested quite low, but would still plan to use organic fertilizer at planting time.

pH. We spread lime in the North and South Main field at 1.5 tons and 1 ton per acre, respectively. Using the Conversions for Amendment Application Rates (page 75), we determined this to be about 75 and 50 pounds per 1,000 square feet (0.75 and 0.5 pound per 100 square feet).

Sulfur. This one is a little harder. We could use elemental sulfur, but that would also cause our pH to drop, which is a problem, since we're trying to raise it. We could use gypsum, which would also add calcium. A little extra calcium is always good as long as we don't throw the Ca:Mg ratio off. However, we're already adding calcium with our lime application. We ultimately decided to use sulfate of potash (langbeinite), since the recommendations call for adding potassium as well. Sulfate of potash also contains magnesium, which we don't need any more of, but we'll be adding a relatively small amount, and the additional calcium from the lime should keep everything in balance.

We do some math and find that we need to apply sulfate of potash at about 140 pounds per acre. Sulfate of potash is about 22 percent sulfur, so that would supply around 31 pounds of it. This comes out to about 3 pounds per 1,000 square feet, or 0.3 pound for 100 square feet. The bag of sulfate of potash says we can apply it at 1 to 2 pounds per 100 square feet, so we know we're in a safe range. (See page 75 for tips on spreading small amounts of a soil amendment over a large area.)

Boron. We decide to use Granubor, an organically certified boron fertilizer. Granubor, which is 14 percent boron, is produced by the same company that makes Borax, an all-natural laundry additive. Borax also works well as a soil amendment, and is easier to source for small-scale growers. We definitely don't want to overdo it with the boron because it can be toxic to plants if the concentration gets too high, so we carefully apply it at 18 pounds per acre or 0.4 pound per 1,000 square feet.

Zinc and manganese. These two micronutrients are not commonly applied as single nutrient amendments. In fact, most growers don't worry about them at all. The good news is that we use kelp meal in our organic fertilizer mix, and kelp contains trace amounts of zinc and manganese. We decide to call that good. Liquid kelp is a great foliar feed, so we may also use that on our plants during the growing season (see page 140 for more information on foliar feeding).

PROBLEM-SOLVING SOIL

The test results we shared and interpreted for you above were specific to that site. Obviously, you'll have to come up with problem-solving solutions for your own garden, based on your own conditions. To help you do that, here's a look at the steps you can take to remedy many common soil composition problems.

YOUR SOIL MANAGEMENT APPROACH SHOULD ALWAYS BE HOLISTIC & BALANCED; AVOID A "ONE NUTRIENT" FOCUS WHEN SOLVING PROBLEMS.

Raised beds and gravel pathways can allow you to grow food on top of a contaminated site.

One thing to be aware of when testing soil is to avoid a "one nutrient" focus when solving problems. Changing one nutrient level in the soil may affect the availability of other nutrients. Thus, your soil management approach should always be holistic and balanced, and should begin by supplying adequate organic matter with compost. Making minor micronutrient adjustments is a futile exercise if you don't have good overall soil structure to start with.

ADJUSTING PH

Soil pH in and of itself does not directly affect plants; it affects the availability of many nutrients, which in turn affects how well plants grow. We want our soil to be between 6.3 and 6.8 to keep nutrient availability in the optimum range for vegetables.

Regular applications of lime can help prevent soil pH from becoming overly acidic.

Add lime to raise pH. There are two kinds of commonly used lime: calcitic and dolomitic. Both types add calcium to the soil; dolomitic lime also supplies extra magnesium. Only use dolomitic lime if you have a magnesium deficiency, or you might cause an artificial calcium deficiency. Calcitic lime is sometimes simply labeled as "agricultural lime" or "garden lime."

+ **Lime application rate:** 50 pounds per 1,000 square feet (1 cup per 10 square feet) will raise pH level by 0.5. One cup equals about half a pound. Avoid using more than double this rate during the course of a season (i.e., don't spread more than 100 pounds per 1,000 square feet in a season).

Add elemental sulfur to lower pH. Elemental sulfur is an organically approved soil amendment that is produced when sulfur impurities are extracted from chemicals during industrial processes. It is often sold in pelleted form, which looks like lentils, and is relatively slow acting in the soil. Depending on temperature and soil conditions, it can take several months to effectively adjust soil pH levels.

+ **Sulfur application rate:** 10 pounds per 1,000 square feet (¼ cup per 12 square feet) will lower pH level by 0.5. One cup equals about half a pound.

To apply lime or sulfur, sprinkle the desired amount over the surface of the soil, and turn it into the top 6 inches of soil with a spade shovel or fork. These amendments take a few months to actually make a difference, so they're best added in fall, after you've cleaned up your summer crops.

DEALING WITH LEAD & ARSENIC CONTAMINATION

High lead levels. Lead is toxic to humans, so it's obviously not something you want in your garden soil. The main problem with lead is not that the plants will absorb it from the soil. Plants will not take up lead unless they're extremely water or nutrient stressed. The risk is that the lead-contaminated dust from the soil will get on your hands and on the crops, and eventually end up in your mouth through these channels.

Lead contamination is usually the result of past industrial activity, or from lead paint on an old house or shed. Lead paint was outlawed in America for household use in 1978, so if you're building a garden adjacent to a building built before this date, you should send in a test for lead before getting started. Similarly, if you know the land has a history of industrial use, you should also send in a test. Not all labs offer lead testing, or offer it only as an add-on to a normal soil test, so check first before sending in a sample.

If your soil test shows high levels of lead, don't worry—you can still have a safe and successful garden. In most situations, it will make sense to build your garden up to avoid contact with the contaminated soil. Build raised beds at least 16 inches deep and fill them with purchased soil. Cover the surrounding paths with a thick layer of mulch to keep dust off of your feet, hands, and garden beds. Wash your crops well before eating, which is a good practice anyway.

Arsenic contamination. Similar to lead, arsenic contamination can be a problem in areas with an industrial history (especially paper mills and smelting), areas that were orchards prior to the 1960s (because of the use of lead arsenate as a pesticide), or underneath structures built with treated lumber manufactured prior to 2003 (decks, play structures, etc.). If you have reason to suspect arsenic contamination, we recommend testing prior to constructing your garden.

Not all labs offer arsenic testing, so check first before sending in a sample. You can also test for arsenic using a home test kit (see Resources, page 292). As with lead, the main problem with arsenic is not that your plants will take it up from the soil, but that the contaminated soil particles will end up on your hands or on plant foliage. If you determine that your soil is contaminated with arsenic, you can take the same steps as those listed above for lead: Grow crops in raised beds with purchased soil, keep contaminated soil near the garden covered with a thick layer of mulch, and wash produce well when harvesting.

SPREADING SMALL QUANTITIES

Spreading amendments at low rates over large spaces can be very difficult. Boron is a great example; application rates are usually only a few pounds per acre. One way to apply small quantities of a given nutrient is to mix it with another amendment.

For example, if you're spreading lime on a 1,000-square-foot plot and need to spread a small amount of sulfate of potash over the same space, weigh out the amount of each amendment you need, mix them together thoroughly, then spread them. A push lawn spreader would be a good piece of equipment for this spreading job.

Alternatively, some amendments can be mixed with water and applied as a soil drench with a watering can or hose-end sprayer. This is only realistic for small and medium-size gardens, but it's a very easy and effective technique. See the Organic Soil Amendments chart (page 78)

CONVERSIONS FOR AMENDMENT APPLICATION RATES

Small-scale production growers sometimes have it rough when it comes to determining how much of an amendment to apply. Many fertilizers and soil amendments have application rates listed in pounds or tons per acre, but most gardeners are dealing with square feet and are applying amendments with measuring cups and watering cans instead of tractor spreaders. Labels are improving as suppliers realize they need to meet the needs of small-scale growers, but here are some conversion rates to help you figure things out:

1 acre = 43,560 square feet

½ acre = 21,780 square feet

¼ acre = 10,890 square feet

⅛ acre = 5,445 square feet

4- by 8-foot raised bed = 32 square feet

4- by 10-foot raised bed = 40 square feet

10- by 10-foot garden space = 100 square feet

1 ton = 2,000 pounds

1 ton/acre = about 45 pounds/1,000 square feet

1 ton/acre = about 4.5 pounds/100 square feet

If you know how many pounds of an amendment you need per acre, multiply it by 0.02 to determine how many pounds you need for 1,000 square feet. Multiply it by 0.002 to determine how many pounds you need for 100 square feet.

1 cubic yard of compost = 1,000–1,600 pounds

1 cubic yard = 27 cubic feet

for soil drench application rates for a variety of amendments. If you happen to have a tractor with a boom sprayer, this technique is highly effective for large acreages as well.

ADJUSTING NUTRIENT LEVELS

PROBLEM	ASSOCIATED PLANT SIGNS	QUICKLY AVAILABLE ORGANIC SOURCES
Low nitrogen	Plants are stunted, slow-growing. Foliage may be yellow or pale green.	Blood meal, fish meal, liquid fish emulsion. Raw manure contains a lot of available nitrogen, but we advise against applying it directly to your soil. Compost it first to avoid potential pathogen risks.
Low phosphorus	Foliage shows a purplish color. Plants grow slowly. Fruiting plants have problems setting flowers and fruit.	Bone meal, commercial high-phosphorus organic liquid fertilizer
Low potassium	Yellowing/dieback of leaf tips (tip burn); unhealthy, stunted fruit.	Sul-Po-Mag/sulfate of potash
Low calcium	Plant grows poorly/is stunted; blossom-end rot in fruits, leaf tip burn.	Foliar feed with liquid kelp, chamomile tea, eggshell tea, commercially available calcium spray (Calcium 25)
Low magnesium	Yellow tinge in leaves that first appears in between veins.	Epsom salts diluted in water and used as a foliar spray
Low boron	Variable symptoms. Hollow stems are characteristic of a boron deficiency in brassica crops.	Borax diluted in water and applied as a soil drench
Low micronutrients/ trace elements	Variable symptoms	

SLOW-RELEASE ORGANIC SOURCES	NOTES
Compost, seed meals (soybean, cottonseed), alfalfa meal, feather meal	
Rock phosphate, colloidal phosphate, composted manures	Potassium is difficult for plants to take up when temperatures are cold. Using a high-phosphorus fertilizer when transplanting helps with new root growth.
Greensand, kelp meal, composted manures, composted wood ash	Potassium can be leached by excessive precipitation or irrigation. This is common in container gardens and in raised beds with limited depth.
Lime (if you also need to raise soil pH), dolomitic lime (if you also need to raise pH and magnesium), or gypsum (if you don't want to alter soil pH, also contains sulfur)	Calcium is very important for the structural health of your crops' fruit and foliage. It must be in proper ratio with magnesium (between 5:1 and 7:1 Ca:Mg).
Dolomitic lime (if you also want to raise pH), Epsom salts (also a good source of sulfur)	Magnesium must be in proper ratio to calcium (between 5:1 and 7:1 Ca:Mg), so don't apply it unless you know you have a deficiency.
Borax, Solubor, Granubor	Apply conservatively; boron can quickly become toxic if overapplied.
Greensand (contains many micronutrients and potassium), kelp (liquid or meal, contains many micronutrients and potassium), Sul-Po-Mag/sulfate of potash (contains sulfur, potassium, and magnesium), elemental sulfur (contains sulfur, also lowers pH), compost (contains many micronutrients), gypsum (contains sulfur and calcium)	Using a blended fertilizer that contains kelp or greensand or using liquid kelp as a foliar spray will help prevent imbalances.

ORGANIC SOIL AMENDMENTS

AMENDMENT	APPLICATION METHOD	NUTRIENT PERCENTAGE
Sul-Po-Mag/ langbeinite/sulfate of potash/K-mag	Mix into soil, use as mixed fertilizer ingredient, soil drench at 4 tablespoons per gallon over 100 square feet.	22% potassium, 22% sulfur, 11% magnesium
Borax (pure sodium tetraborate)	Mix into soil, soil drench at ¼ teaspoon per gallon and spread over 30 square feet.	About 11% boron
Gypsum	Mix into soil.	22% calcium, 12% sulfur
Epsom salts	Mix into soil, foliar feed or soil drench at 4 tablespoons per gallon of water.	10% magnesium, 13% sulfur
Greensand	Mix into soil, improves structure.	3% potassium, trace minerals
Kelp meal	Mix into soil, use as mixed fertilizer ingredient.	1% nitrogen, 2% potassium (1:0:2), many trace minerals
Blood meal	Mix into soil, use as mixed fertilizer ingredient.	12% nitrogen (12:0:0)
Rock phosphate	Mix into soil.	3% soluble phosphorus (0:3:0), 20% calcium, 20% insoluble phosphorus
Bone meal	Mix into soil, use as mixed fertilizer ingredient.	2% nitrogen, 14% phosphorus (2:14:0), up to 24% calcium
Calcitic lime	Mix into soil.	Approximately 30% calcium, 5% or less magnesium
Dolomitic lime	Mix into soil.	Less than 30% calcium, greater than 5% magnesium

CHAPTER FIVE

BUILDING & MAINTAINING SOIL QUALITY

The quality of produce and yield from your garden is dependent on the quality of the soil. Because of this, you should do everything possible to make garden soil the best it can be right from the start, and then continue to work it year after year to maintain and improve its quality. This is necessary for any gardener expecting high production and multiple harvests from a single plot of soil.

Most existing soils, especially in a residential yard or urban plot, will not have enough available minerals or nutrients to grow high-yielding vegetable plants. Thus, it's vital that you take steps to add amendments when building your garden, and continually maintain them over time. If you're doing a good job, the soil quality should improve with each gardening season.

Buckwheat is a fast-growing cover crop that can be used to quickly improve the quality of your soil.

first priority: build up humus

 Humus—what's created when soil micro-organisms break down raw organic matter—is an essential component for good garden soil. It acts as a glue to hold soil aggregates together and creates what growers call tilth, or good soil structure. Humus gives garden soil a wonderful loose, crumbly feeling and a texture that doesn't completely collapse into individual particles when you work it.

Humus also acts like a sponge in your soil; it helps maintain proper moisture levels. In soil without humus, water will either leach away too quickly for plants to use (very sandy soil), or will make the soil waterlogged and difficult to work for long periods (very clayey soil). Humus-rich soil will take up water and hold it right where your crops can use it. As humus continues to break down, it releases nutrients that your crops can use to grow. Amazing stuff, that humus.

JUST ADD ORGANIC MATTER

Because humus breaks down, its presence can only be maintained by adding organic matter to the garden on a regular basis. The easiest and fastest way to do this is to add compost. Because compost is already partially decomposed, it's easily converted into humus. Also, the nutrients in compost have been stabilized, so they're less likely to leach from the soil than from raw organic matter. Good compost also contains beneficial microorganisms, helping to keep the biotic community of your soil in good health.

It's important to note, however, that not all compost is created equal. You might hear a recommendation such as "add X amount of compost to your garden each year, and you'll be able to grow wonderful vegetables without adding any other fertilizers." This might be true in climates with warm growing seasons and the highest-quality compost, but it does not hold true in every situation.

For example, compost made with horse manure will supply a different amount of nutrients than compost made with chicken manure or compost made from plant material. Finished compost will begin to supply nutrients quickly, but compost that's still breaking down may actually tie up nitrogen for a period of time because the soil bacteria are using it to process the organic matter.

FOUR WAYS TO BOOST YOUR SOIL

There are four simple things you can do to maintain your soil quality from one year to the next:

1   Add a 1- to 2-inch layer of compost over your garden once a year.

2   Monitor and maintain proper soil pH.

3   Use additional organic fertilizer every season to ensure proper nutrient levels.

4   Protect your soil with mulch or cover crops, especially during winter.

Cover crops like buckwheat can be broadcast seeded.

Compost is an essential part of any organic soil management plan and should be applied at least once a year.

making quality compost

Compost is a key ingredient for your soil fertility, and you have two choices for how to obtain it: Make your own or buy it. Composting at home is an excellent way to deal with garden waste in an environmentally friendly way. When you make your own compost, you theoretically have control over the quality of the compost that goes back into your garden. However, making compost can be a lot of work, and it takes some skill to make really good compost. In addition, many gardeners struggle with sourcing enough bulk ingredients like manure or straw to keep their nutrients in balance. Many growers do

both—they buy large quantities of compost to get their garden started and for major fertility overhauls, and also create a pile to handle garden waste and produce smaller amounts for top-dressing beds.

COMPOST PILES & WINDROWS

You can produce a surprising amount of compost by maintaining catchall piles. This is probably what you think of when someone mentions composting; it's the process of making a mound of your kitchen and garden scraps, yard waste, and

CARBON-TO-NITROGEN RATIOS OF COMPOST INGREDIENTS

BROWNS

INGREDIENT	C:N RATIO	NOTES
Branches	500:1	Use small-diameter branches and twigs or chip them before adding.
Dried grass cuttings	50:1	
Dried deciduous tree leaves	70:1	
Sawdust	400:1	
Shredded cardboard	350:1	
Shredded newspaper	170:1	Make sure paper has natural-based inks such as soy.
Straw	75:1	
Wood ashes	25:1	
Wood chips	400:1	

GREENS

INGREDIENT	C:N RATIO	NOTES
Coffee grounds	20:1	
Freshly cut grass	15:1	
Garden waste	30:1	
Kitchen waste	30:1	
Manure	20:1–40:1	Ratio will vary depending on animal source and freshness.
Peat	70:1	
Weeds	30:1	Do not compost weeds with seed heads or very aggressive weeds.

HOW TO MAKE COMPOST

If you have the space, time, and access to ample feedstocks, making your own compost is a great way to feed your garden and process your plant wastes.

GETTING THE RIGHT RATIO

When adding materials to the compost pile, it's important to get the right ratio of carbon ("brown" debris) to nitrogen ("green" waste). Your goal is to achieve a carbon-to-nitrogen ratio of about 25:1. Bacteria will break down organic waste into usable compost most efficiently at this ratio.

Certain ingredients lend themselves incredibly well to this process. For example, the C:N ratio of horse manure is about 30:1 to 35:1, so horse manure on its own, or with a little bit of straw bedding mixed in, makes amazing compost. If there's too much straw bedding or the manure is mixed with sawdust or wood chips, you'll probably need to add more high-nitrogen materials. Chicken manure mixed with bedding makes great compost, as does young grass clippings or coffee grounds mixed with brown leaves or straw.

You don't have to be too precise with your ratios or measuring amounts of ingredients when building a compost pile. Many home growers are limited to the ingredients that have come from their yard, garden, kitchen, and other nearby sources, so may not have the luxury of using an ideal balance of ingredients.

After gaining some experience maintaining compost piles, you'll develop a good general feeling for whether a given pile needs more green or brown material. It's a good idea to keep a straw bale or stash of dead leaves on hand to mix in with green, high-nitrogen garden waste or kitchen scraps. When the pile reaches the desired size—about 3 feet by 3 feet by 3 feet—you should start turning it. If you run into problems, add more browns or greens each time you turn it to balance things out.

WHAT *NOT* TO INCLUDE

We recommend not including vigorous weeds and those that have set mature seed in your home compost pile. This is because it can be difficult to create temperatures high enough to kill weed seeds and invasive weed roots. The same is true with diseased or heavily insect-infested plant matter (see page 159 for how to deal with this situation). Also, be wary of using sawdust, bark mulch, or wood chips in your compost, as the high carbon levels in these materials can tie up soil nitrogen if the compost is not fully finished when applied. Sawdust is generally devoid of other nutrients and minerals, so compost made with large quantities of sawdust tends to be of lower quality. Other ingredients to avoid: meat, bones, colored paper, waxed cardboard, human waste, oils and fats, pet waste, and pine needles.

BIGGER IS BETTER

Composting works better when you go big—it's easier to build up the heat that helps drive the composting process in a larger pile. If you're composting in a catchall pile, 3 feet by 3 feet by 3 feet should be the minimum volume of the pile before you start turning it. Smaller piles break down very slowly, and won't ever develop enough heat to kill pathogens and weed seeds.

The compost pile should also be consistently moist, but not overly wet—like a wrung-out sponge—to facilitate decomposition. Water the pile with a hose if it feels too dry. Cover the pile with a tarp during heavy rains so it doesn't get soaked and to prevent nutrients from leaching out of the pile. If the pile does get too wet, you can add dry brown material to help soak it up.

TURN, TURN, TURN

Turn the pile frequently with a pitchfork or garden fork—as often as once a week—for rapid breakdown. You can turn it less frequently (every other week, once a month, or less often), but it will break down more slowly. If you're concerned about rodents and other critters, turn the pile at least every 2 weeks. This maintains high temperatures in the pile and greatly speeds breakdown, both of which make the pile less attractive to pests.

When turning a pile, you want outer layers to move to the middle of the pile, and the middle of the pile to end up on the outside. To do this, start by forking the top of the pile onto the ground next to the pile. Keep moving the pile over from the top down until you have a new pile. The

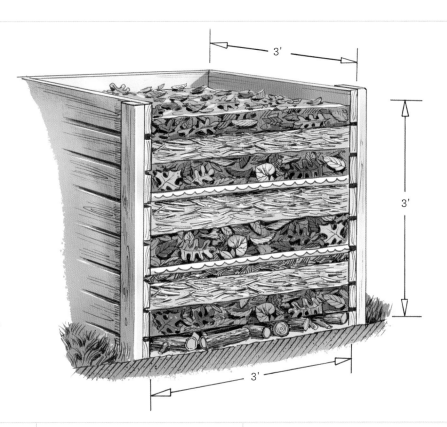

IF YOU'RE COMPOSTING IN A CATCHALL PILE, 3 FEET BY 3 FEET BY 3 FEET SHOULD BE THE MINIMUM VOLUME OF THE PILE BEFORE YOU START TURNING IT.

top of the old pile is now at the bottom of the new pile, and vice versa. If you have space, you can leave the new pile where it is. If not, you can fork it back to its original position.

TRACKING THE TEMPERATURE

If you have a soil thermometer, you can use it to help with monitoring the pile. Stick it in the pile after you turn it for the first time. Check the temperature each day. Over the course of a few days or a few weeks, depending on your balance of feedstocks, the temperature will rise, peak, and then start to fall. If it starts to fall before reaching 130 to 140°F (54 to 60°C), add more nitrogen-rich ingredients the next time you turn it. If it gets over 150°F (65°C), it probably needs more carbon-rich ingredients. Turn the pile as soon as possible, as

temperatures over 150°F (65°C) can kill beneficial bacteria.

When ambient temperatures are low, especially when they're below freezing, the composting process usually slows or stops. Home compost is most easily made during late spring, summer, and early fall in most climates. During winter, you can cover the pile with a tarp and let it sit until more favorable temperatures return in spring.

FINE-TUNING YOUR PILE

If your compost pile smells particularly strong, bad, or like ammonia, add more brown materials. If you accidentally get too many brown materials in your pile (or just have a lot of them and want to get them to compost), you can add blood meal or another high-nitrogen organic fertilizer to balance out the C:N ratio. Consider covering your pile with a thin layer of

brown materials, such as straw or leaves, to minimize insect pests and odors.

The compost is finished when it looks like soil and feels and smells earthy. A compost pile that is well balanced and turned frequently might be ready to apply to the garden in 3 months. A less frequently turned pile might take 6 months to a year.

If you have space, make compost in separate batches. When one pile has enough bulk to start turning, stop adding new materials (unless balancing the C:N ratio is necessary). Start a second pile while you're turning the first one, and a third pile while you're turning the second. This allows all the ingredients in each pile to break down fully and uniformly. If you add a lot of fresh materials to a pile midway through the composting process, they won't be decomposed when the rest of the pile is finished.

other organic materials, and turning it frequently to mix oxygen in. If you maintain a proper carbon-to-nitrogen ratio (see page 83), keep the pile evenly moist, and turn it regularly, you'll be able to produce quality compost.

If you're able to source large quantities of raw materials and want to make a lot of compost, make a windrow instead of a pile. This is basically a long row of composting materials that's at least 3 feet tall at the peak, 6 to 7 feet wide, and as long as you want to make it. The windrow shape is easier to turn than a single massive pile. You can turn a windrow by hand, but most growers use a small tractor with a front-end loader or a specialized compost turner. If you have a manure spreader for your tractor, you can use it to build the pile. Fill the spreader with raw composting ingredients, set the spreader to unload as quickly as possible, and drive forward as slowly as possible. The spreader will unload the ingredients into a windrow shape. To get the right shape, you'll need to fine-tune this method depending on your tractor and spreader. For suggestions on where to find more information on making large quantities of compost, see Resources (page 292).

STATIC, AERATED COMPOSTING

Don't want to spend time turning your compost? Some growers construct homemade aerators to force air through their compost piles using blower fans and perforated pipes. This supplies enough oxygen for the pile to heat up and break down without turning. A search of the internet for "static aerated composting" will yield small-scale designs for such a system. You can also "cold compost," which is basically piling up your compost ingredients and letting them sit until they've broken down. The risk with this method is that the pile doesn't heat up as much as an aerated pile does, so weed seeds and pathogens may still be present in the pile. A "cold" pile might take anywhere from 2 to 5 years to be ready for use in the garden.

SOURCING BULK MATERIALS TO COMPOST

Whichever compost technique you choose, you may find yourself wishing you had more stuff to throw on the pile, especially once you've seen how beneficial compost can be for your crops. Here are some tips on acquiring more material.

+ Many grocery stores will give you damaged or old produce for free. This makes a great high-nitrogen addition to compost piles. It helps if you bring your own containers (5-gallon buckets work well) and show up at a set time that's convenient for the produce manager.

+ Coffee shops and restaurants will often let you have their spent grounds, another nitrogen-rich compost feedstock. Ask if you can leave 5-gallon buckets for them to fill up.

+ Check with neighbors who rake leaves in their yard. They'll probably be happy to hook you up with a few trash bagfuls. Hardwood leaves are a great carbon source and are loaded with minerals. Return the favor with a few bags of salad greens in spring.

+ Horse stables are often a great source for free manure. Manure mixed with straw bedding is easiest to compost. If the horses are bedded on sawdust or wood chips, add additional nitrogen to the pile and compost for a full year before you add the finished product to your garden.

+ Chicken, pig, and cow manure from a local farm are all great nitrogen sources for your compost pile.

+ Keeping your own livestock is a great way to convert garden waste into manure for compost. Even urban dwellers with relatively small lots can successfully raise a happy flock of chickens (see Resources, page 292, for more information).

Compost can often be purchased in bulk and applied to the garden by wheelbarrow or bucket.

purchasing compost

↓   Making your own compost can be a great way to ensure a high-quality product, but it is something we see many growers struggle with, especially in urban environments. If you don't have access to quality feedstocks for your compost and a way to transport them (e.g., a truck), it can be difficult to build a top-notch pile. If this is the case for you, don't worry. There's nothing wrong with buying compost. Even if it's only moderate quality, you can still have a great garden if you use organic fertilizers and soil amendments appropriately. And if your compost ends up being subpar, you can always try a different compost supplier next season.

Ideally, you'll be able to find a locally made product created specifically for vegetable gardens. Avoid compost that is sold for use on ornamental beds as it's likely to contain lots of high-carbon materials, like undecomposed wood and bark, which can reduce the availability of nutrients for your plants.

Obviously, a certified organic product will help ensure quality ingredients, but well-made compost does not need a certification to be a great product. Talk to the provider and learn how the product is made and what it is made from before loading it up.

Generally speaking, compost made with animal manure will provide more usable nitrogen, phosphorus, and potassium. Compost made strictly with plant waste is still a great source of organic matter, but will yield less of these nutrients. Here are a few additional items to consider when purchasing compost.

THE IMPACT OF CLIMATE ON COMPOST IN THE SOIL

Believe it or not, the climate you live in affects the ability of your compost to supply nutrients to your crops. Air temperature affects soil temperature, and soil temperature affects how quickly soil microorganisms can break down organic matter and make nutrients available to plants. The warmer the temperature, the faster the organisms will work. Thus, if you're growing in a warmer climate, you can expect to get more usable nitrogen, phosphorus, and potassium from your compost than you would in a cooler climate. In this case, you'll need less organic fertilizer.

This also means that your compost and humus will break down more quickly, and will need to be replenished more frequently. Phosphorus is notoriously unavailable in cooler temperatures, so using supplemental organic fertilizer is particularly important early in the season.

WATCH OUT FOR CONTAMINANTS

Herbicides. Compost can become contaminated with persistent herbicides if they're present on the compost feedstock. These pesticides can damage crops in your garden. This problem is rare, but it's worth checking to see if your supplier screens for it.

Arsenic. Some confinement chicken operations use arsenic in their feed, which can contaminate chicken manure compost. Again, ask your supplier if they screen for this.

Heavy metals. In the past, composted biosolids (human waste) often contained heavy metals that you wouldn't want in your soil. Today, sewage systems are more carefully regulated and have become much cleaner. Composted biosolids can be a clean, viable option for food production.

Such compost must be made to exacting standards to ensure that all pathogens have been killed, and should be tested for heavy metal content. If your supplier can't speak to these practices, avoid composted biosolids. Uncomposted sewage sludge should not be added to an edible garden under any circumstances.

WOOD ISN'T GOOD

Any compost made with large quantities of wood chips or sawdust is not suitable for vegetable production. In the short term, composts or other soil amendments that contain high levels of carbon—such as wood chips, bark, or sawdust—will actually decrease nutrient availability in the soil. It can, however, be great for mulching around trees and perennials. Many landscaping companies and nurseries carry compost made with woody materials; make sure you don't get the wrong product if you're buying from these sources. These materials may be labeled as compost, but are more accurately described as mulch. The first year we ran our business, we unknowingly built all of our gardens using purchased compost with a lot of sawdust in it. Every crop in every garden grew poorly until we tested the compost and found out what was happening—no nitrogen!

NOT HOT

Compost that is hot, steaming, or has a strong smell is probably not fully finished and could "burn" plants, either physically or with too many nutrients. Fully finished compost will have an earthy, soil-like smell and feel. If you happen to buy unfinished compost, you can pile it up and turn it a few times until it breaks down and cools off. It's also okay to add directly to the garden, but you'll want to wait at least a few weeks to plant into these areas. We like adding compost to the garden in fall to make sure it's fully broken down before we begin planting in spring.

improve the soil with cover crops

↓ Cover cropping is the practice of growing a crop to generate organic matter, protect the soil, or capture nutrients. Cover crops are often planted in late summer and fall, so that the crop is mature enough to offer good protection for winter. Cover crops can also be a great midsummer soil management option between a spring and fall crop. Or they can protect the soil if you take a bed out of production for the year to let the soil rest or to break up pest and disease cycles. Most growers use a combination of a leguminous cover crop and a grass. The legume fixes nitrogen from the air and adds it to the soil, and the grass supplies carbon to balance the nitrogen from the legume.

This cover crop of field peas is flowering and ready to be mowed and turned in.

USE COVER CROPS TO PROTECT THE SOIL IN WINTER

For winter protection, cover crops are the best thing you can do for the soil. As they grow, the crops take up excess nutrients left in the soil and prevent them from leaching away. The roots provide a haven for beneficial fungi throughout the winter. When you mow the cover crops and turn them into the soil, they release usable nutrients and help build organic matter.

For winter cover, you can choose between cover crops that winter kill—meaning they die at a certain temperature—or those that overwinter and grow through spring. Winter-kill cover crops are great for preceding early-spring crops like broccoli or kale because they're much easier to turn under in spring. Overwintering cover crops are best for later-planted crops like tomatoes because they continue to grow and generate biomass in spring. It's important to know your climate when choosing, because a given crop might overwinter in one area and winter kill in another.

It's also possible to leave the cover crop or its residue in place on top of your beds as a form of mulch. If you plant a cover crop that will winter kill, consider leaving the residue on the beds come spring and transplanting your crops into holes in the decomposing matter. This method is best suited for beds where the crops will be transplanted as opposed to direct seeded into the beds.

START SMALL

Cover crops do have drawbacks for the small-scale production gardener. They take quite a bit of work to mow and turn in, especially if you're only using hand tools. They can also tie up valuable production space in spring, because it takes 3 to 5 weeks for the crops to break down after they've been turned in. If the majority of your garden is planted in cover crops, you may not have

COVER CROPS AS LIVING MULCH

You can sow a cover crop as a living mulch to grow alongside your primary crops, shading the soil from drying sunlight and suppressing weeds. Living mulches can work well, especially when used around tall, vigorous crops. Keep in mind that allowing a living mulch to set seed will establish it as a permanent fixture in the garden (a.k.a. a *weed*). Living mulches need to be mowed regularly to prevent this from happening. Similar to other cover crop applications, we recommend starting small and experimenting to see if this technique works for you.

enough room to plant early-season vegetables in a timely fashion.

Because of these drawbacks, we recommend starting small with cover crops. Don't plant more than 100 square feet in cover crops the first time you try them, especially if you're using hand tools to turn them in, and don't plant more than a quarter of your garden space with them. Once you get a feel for managing cover crops, you can adjust your square footage to match your needs.

HOW & WHEN TO SOW COVER CROPS

Start by clearing old crops and raking the soil smooth. You don't need to work the soil deeply unless it's become compacted for some reason. Once you have the seedbed established, you can sprinkle smaller-seeded cover crops (clover, rye, oats, vetch) over the surface of the soil. Use a rake to gently pull soil over the seeds, then water. Keep the cover crops watered until they emerge.

If you're planting larger-seeded cover crops such as field peas, you'll need to create furrows so the seeds can be buried a little deeper. On a small scale, use a rake or hoe to make 1-inch-deep furrows 6 inches apart. Sprinkle the field peas into the furrows, then rake soil over the furrows. Aim for about 8 seeds per foot, but don't worry about being too precise. If you're sowing rye or oats with field peas, start by making the furrows. Spread the peas, then sprinkle the rye or oats over the soil. Finally, rake it all in and water well. Keep the cover crops watered until they emerge.

Cover crops can also be sown beneath mature crops. By sowing while the previous crop is still in place, you can establish the cover crop earlier in the season, allowing it to mature while the weather is still good for quick growth. Undersowing can also lead to better germination rates since the seeds benefit from the shade and protection of the existing crops, and don't dry out or get snatched out of the soil by birds.

TURNING THEM IN

Plan to turn in your cover crop at least a month before you want to plant vegetables. If the cover crop hasn't winter killed, you'll need to mow it first. On a very small scale, you can do this with a pair of long-bladed garden shears. For larger spaces, you'll want to use either a string trimmer or a mulching lawn mower that drops the trimmings back onto the soil.

After mowing, work the cover crop into the soil. Use a spade shovel or spading fork to uproot the crops, turn them over, and chop clumps into pieces. Plan to check the bed once a week, and further turn and chop any crops that are starting to resprout; this can be a problem in areas with rainy spring weather. A walk-behind rototiller makes this task much easier. We don't advocate habitual use of rototillers to prepare beds because overusing them can create a hardpan, but they do an excellent job of working in cover crops efficiently.

If your climate is very dry, you may need to water the bed to promote decomposition. After about 4 to 5 weeks, the beds should be relatively clear of debris and visible plant matter. If there are still lots of green/living bits of the cover crop, continue to work the soil until they've broken down.

If you're planning on adding compost to the soil, you can use it to help kill your cover crop. After you mow and turn in the cover crop, spread the compost over it. The layer of compost helps smother the cover crop and makes sure it can't photosynthesize. Mix this compost into the soil once the cover crop has decomposed and you're ready to prepare the planting bed.

COVER CROPS FOR EVERY GARDEN

Many plant species can serve as cover crops. Your choice will vary based on your cover cropping goals, soil conditions, and the time of year. Some cover crops, such as clover and peas, are leguminous, which fix nitrogen from the air. Some, such as annual rye, create an allelopathic effect in the soil that helps prevent weeds from germinating. Some cover crops germinate in cold weather, which makes them a good choice for early- and late-season plantings; others are very frost sensitive and grow best midsummer. It should be possible to identify a cover crop that meets your needs for virtually any situation.

Buckwheat. This crop grows very rapidly and has a short life span, so is useful as a midseason cover crop between spring and fall plantings. It's quite tender, so will winter kill in most of North America. It doesn't germinate well in cool soil, so it has limited applications for winter cover.

Clovers. These are leguminous cover crops that are great for creating a long-term pasture (except crimson clover, see page 92). We don't use clovers very much in annual gardens because most types take too long to generate enough root and top growth to create organic matter and protect your soil. They also don't grow quickly enough if sown in fall to effectively protect the soil for winter. If you have space to grow clover for a season,

it's great for breaking up compaction and fixing nitrogen, and is an excellent forage for chickens. For all clovers, be sure to mow before seed set to prevent clover from becoming a weed problem; turn in when desired.

Field peas. Field peas are a more manageable legume than vetch because they generate less top growth to turn in. Field peas winter kill in many parts of the United States, but overwinter well in the Pacific Northwest. An added benefit is that you can harvest and eat the growing tips when the plant starts to mature—delicious! Field peas are often sown together with oats.

Hairy vetch. This vigorous legume grows well in fall for winter cover and generates a huge amount of soil nitrogen for the following year's crop. Because of all the top growth, vetch can take a bit more work to turn into the soil in spring. Rye and hairy vetch make a great combination for winter cover.

Mustard greens. Research has shown that mustard greens can greatly reduce soil-borne diseases when grown as a cover crop, so you can use mustard greens as a "fall cleaning" cover crop in a greenhouse or field that has disease issues.

Oats. Oats are another commonly used grass. They are often grown in combination with field peas for a great winter-kill cover crop in most of the United States.

Winter rye. When purchasing rye, make sure to get the species *Secale cereale*. There are other grasses known as ryegrass (such as *Lolium multiflorum*), which have very different growth habits. Winter rye is one of the most commonly used grasses for a cover crop. It's extremely hardy, and it germinates well in cold soil. Rye requires vernalization to set seed, so it needs to be planted in late summer/fall if you want it to produce grain or straw.

IO COMMON COVER CROPS

CROP	WHEN TO SEED	MANAGEMENT	WINTER KILL TEMP. (°F)	SEEDING RATE	ALTERNATE SEEDING RATE	NOTES
Buckwheat	Spring–early summer	Mow at flowering and turn in.	32	2–3 pounds/1,000 square feet	8–12 cups/1,000 square feet	
Clover, red	Spring	Mow when flowering to prevent seed set; turn in 4–5 weeks before planting crops.	–20	0.5–1 pound/1,000 square feet, alone or with grass or grain	2–3 cups/1,000 square feet	Most cold-tolerant clover.
Clover, New Zealand white	Spring	Mow when flowering to prevent seed set; turn in 4–5 weeks before planting crops.	–20	0.25–0.5 pounds/1,000 square feet, alone or with grass or grain	1–2 cups/1,000 square feet	Longest-lived clover; good for permanent pastures and areas with foot traffic.
Clover, crimson	Late summer	Mow and turn in the following spring.	0	0.5–1 pound/1,000 square feet, alone or with grass or grain	2–3 cups/1,000 square feet	Usually grown as a winter annual cover crop.
Clover, sweet	Spring–early summer	Mow when flowering to prevent seed set; turn in 4–5 weeks before planting crops.	–20	0.5–1 pound/1,000 square feet, alone or with grass or grain	2–3 cups/1,000 square feet	Especially good at renovating worn-out soil.
Peas, field	Spring for mid-season cover, late summer for winter cover	Mow and turn in when desired; grown as a winter-kill cover in cold climates.	5–10	5 pounds/1,000 square feet alone, 3 pounds/1,000 square feet when sown with grass or grain	6–8 seeds per foot in furrows 6" apart	

CROP	WHEN TO SEED	MANAGEMENT	WINTER KILL TEMP. (°F)	SEEDING RATE	ALTERNATE SEEDING RATE	NOTES
Vetch, hairy	Spring for mid-season cover, late summer–fall for winter cover	Mow at or before flowering to prevent reseeding; turn in when desired.	−20	1 pound/ 1,000 square feet with rye or oats	5 cups/ 1,000 square feet	
Mustard greens	Early fall	Mow at flowering and turn in immediately for best disease suppression.	5–10	0.25 pound/ 1,000 square feet	1 cup/1,000 square feet	
Oats	Spring for mid-season cover, late summer for winter cover	Mow and turn in when desired; grown as a winter-kill cover in cold climates.	20	4 pounds/ 1,000 square feet	9 cups/ 1,000 square feet	
Rye, winter	Spring–early summer for mid-season cover, late summer to mid fall for winter cover	Mow and turn in when desired; grown as a winter-kill cover in cold climates.	−35	4–6 pounds/ 1,000 square feet	9–13.5 cups/ 1,000 square feet	

mulching for winter

↓   It's important to protect your soil in winter, either with cover crops or with mulch. During winter, periods of high precipitation can leach nutrients and acidify the soil. Leaving the soil bare can significantly decrease your population of beneficial mycorrhizal fungi. Covering your soil with an organic mulch in winter provides a physical barrier to leaching. Mulching is much less management intensive than growing cover crops. The main drawback to mulching compared with cover cropping is that it does not leave living roots in the soil, which is the best way to protect the beneficial mycorrhizal fungi. Here are a few options for winter mulches.

Compost. As with soil, nutrients can be leached from compost during winter; however, compost does provide soil protection and creates an easy-to-work-with soil surface for the next season.

Shredded hardwood leaves. Can be turned directly into the soil in spring to add organic matter and nutrients.

Grass clippings. When used as winter mulch, these can be turned into the soil in spring.

Hardwood leaves. Use a thick layer for winter protection (4 to 6 inches). Avoid turning into the soil in spring so as not to tie up nitrogen; instead, rake off the remaining mulch and compost.

Rolled burlap or burlap sacks. Don't turn these into the soil! Instead, store and reuse next year. These also work for midseason mulch, but can be difficult to lay in between plants.

Straw. Use a thick layer (4 to 6 inches). Avoid turning into the soil in spring so as not to tie up nitrogen; instead, rake off the remaining mulch and compost or reapply when crops begin to mature. Whether or not your straw supplier is certified organic, it's important to verify that the straw you're buying was not sprayed with chemical herbicides. Chemical herbicides can persist for some time, and unwittingly adding them to your garden via straw may lead to years of problems.

Straw is an inexpensive and effective way to protect the soil in winter or to suppress weeds and conserve soil moisture during the growing season.

start a soil management calendar

↓ There are any number of ways to manage your soil—and just as many ways to keep track of your soil management tasks. Below is a sample calendar that can be used as a general outline for your own plan.

LATE SUMMER

+ Test soil.

+ Begin sowing cover crops, if they're part of your management plan.

FALL

+ Turn in lime or sulfur to balance pH, based on soil test.

+ Add compost to all beds.

+ Mulch or continue sowing cover crops for winter soil protection.

WINTER

+ Update planting plan to ensure proper rotations.

+ Order cover cropping seeds.

+ Revisit any fertility problems you had over the past season; research appropriate amendments to deal with deficiencies.

+ Relax/take a nap/read old Garfield comics.

SPRING

+ Turn in cover crops/prepare beds.

+ Mix in organic fertilizer at planting time.

SUMMER

+ Keep planting.

+ Sow midseason cover crops to precede fall plantings, if desired.

+ Side-dress/foliar feed heavy-feeding crops.

+ Mulch crops as appropriate.

CHAPTER SIX

SUPPRESSING THE COMPETITION

Your garden soil will have weeds. Even if you source weed-free materials, the wind will blow weed seeds into your garden starting the first day you build your beds. Weeds are a problem for your garden soil because they absorb nutrients that you want available for your crops. The key is to get on top of weeds from the outset and never to let them go to seed.

Weeding will always be a part of organic garden maintenance. Your goal is simply to reduce the overall production of weeds in your garden. This strategy has two main benefits: It reduces your crop's need to compete for nutrients, light, and moisture; and it reduces the amount of time you spend weeding, so you can focus on more important tasks like planting, fertilizing, and harvesting. It also has the benefit of making the garden look more attractive.

Regular soil cultivation ensures weeds are easy to remove and never have a chance to set seed. The hori-hori knife is a great hand tool to clear weeds big and small.

develop a weed-reduction strategy

↓ A weed is any plant that is growing in the wrong place, whether it's a dandelion or a self-sown tomato volunteer. Some of these plants are edible and can create an additional harvesting opportunity (see page 250 for a list of edible weeds), but they can also crowd out more desirable crops and become a nuisance if allowed to spread without control.

IDENTIFY & PRIORITIZE

Weed seeds can remain dormant in the soil for many years, waiting for the right conditions to germinate and grow. Freshly dug soil with regular summer watering might just be the opportunity they've been looking for. If you've just created a new bed, don't be surprised if you start noticing weeds that you've never seen before. Learn to identify the unknown plants and prioritize their removal based on their aggressiveness.

WEED EARLY, WEED OFTEN

Every garden has weeds, but diligently clearing them from the garden throughout the season will keep competition with crops to a minimum. The

key to weed management is to start early and work often. Almost all weeds are easy to eliminate with minimal effort when they're small. Regular, weekly weeding sessions will prevent weeds from taking hold and, more importantly, will keep them from maturing and setting seeds. A flowering weed is a ticking time bomb and should be removed from the garden immediately. At a minimum, the flower stalk should be cut down. If you stay on top of weeds and don't let them set seed, your weed issues will be less and less each year.

CHOOSE THE RIGHT TOOL

Smaller gardens can be effectively weeded by hand with a trowel or hori-hori knife. We recommend keeping one of these with you at all times in the garden, so that you can dig out weeds as you encounter them. Every little bit makes a difference. If your garden is larger, you'll want to use a stand-up weeding tool such as a hoe.

WEEDING EFFICIENTLY WITH A HOE

Using a hula, wire, or collinear hoe will allow you to weed larger areas much more efficiently than using a hand tool. If your garden is very large and has long straight rows, consider using a wheel hoe. Cultivating small weeds with a hoe is best done on a hot, dry day because it takes very little disturbance to kill them. If you're digging up larger weeds with a trowel or shovel, they'll usually come out more easily if the soil is moist, so wait until after a rain or irrigation session.

With any of these tools, getting out and using them once or twice a week should keep problems at bay in most gardens. Take the time to get comfortable with your tool and move through the garden intentionally so as to not damage your crops. Hoes should be kept sharp and clean. The tool is intended to cut smoothly through the soil, keeping your work as efficient and effective as possible and disturbing your crops as little as necessary. Invest in a sharpening stone; a few

minutes of tool care will save countless hours of work out in the garden.

FLAME WEEDING FOR MINIMAL SOIL DISTURBANCE

A flame weeder is simply a long steel torch attached to a propane tank. Once lit, the torch can be waved over the surface of the soil to singe and kill weeds. Small flame weeders are relatively inexpensive and can be effective in suppressing especially thick patches of small broad-leaved weeds. They're less effective with mature weeds or weeds with significant root structures, such as perennial grass. Another benefit of flame weeding is that, since the process doesn't disturb the soil, it doesn't bring new weed seeds to the surface.

It's not necessary to actually ignite and burn your weeds with a flame weeder; a brief burst of heat just long enough to make the weed droop slightly is sufficient to kill broad-leaved weeds. The best practice is to move the torch across the bed, keeping the torch 4 to 6 inches above the soil, at about the pace of a slow walk. When you go back to inspect, the weeds will have wilted and/or turned brown.

PRE-EMERGENT FLAME WEEDING

One effective technique is known as pre-emergent flame weeding. The goal is to direct seed a crop and then flame weed over the top of it just before the crop emerges. Any young weeds will be killed, and your crop will emerge in a clean bed. Pre-emergent flame weeding is most commonly done with beets and carrots, but it can work on almost any direct-seeded crop if it's timed properly. Here are a couple of tips.

+ Cover the end of your direct-seeded row with row cover, plastic, a glass panel, or a cold frame. The seeds under the cover should germinate a day or two earlier than the rest of the seeds in the bed. Flame weed the entire bed surface as soon as you see the seeds under the cover emerge.

+ Sow faster-germinating crops at the end of the row as indicators. For example, sow a row foot of beets at the end of a bed of carrots and flame weed once the beets have emerged but before the carrots pop up. Similarly, you can sow radishes at the end of a row of beets.

A FEW WORDS OF CAUTION

Flame weeding is a technique that might take a little practice to perfect, and for which safe practices must be followed. Flame weeding is best done on a hot, dry day with little or no wind. Always wear closed-toe shoes or boots and non-flammable long pants when flame weeding, and stand upwind from the flame. Be aware of dry vegetation and other fire hazards near your garden, and don't flame weed when wildfire danger is high or if a burn ban is in effect. Keep a hose handy in case of an emergency. Always read and follow the manufacturer's directions for your flame weeder.

Flame weeding can create a stale seedbed immediately prior to planting.

TARPING TO MINIMIZE WEEDS

Another effective technique for weed suppression involves using an opaque, dark-colored tarp to smother weeds. This method (called occultation) has been repopularized in recent years. Simply place a heavy tarp over your bed, and leave it in place for an extended period. Tarping times will vary based on the type of weeds you are combating and the temperature. As a rule, plan for 3 to 4 weeks in summer and 6 or more months in fall and winter. Be sure to use a UV-stabilized material so there's not tattered bits of tarp littering your garden in fall. Five- to 6-mil silage tarps work well (see Resources, page 292, for suppliers). Used EPDM pond liner is also very effective, but only suitable for small spaces because it is so heavy. Make sure to weigh the sides of your tarp down with rocks, soil, or sandbags every few feet. Tarps love to catch the wind, so inadequate weights will ruin your well-laid plans.

The tarp creates a warm, moist environment that encourages weed seeds to germinate. When the weeds emerge, they smother and die from lack of sunlight and air. When you remove the tarp, you'll have a weed-free bed with good soil moisture. Tarping is also highly effective at killing and helping to break down cover crops. Like any method, however, it has some drawbacks. Rodents love to burrow and make nests under tarps, so if you have a rodent problem this might not be the best choice for you.

DON'T FORGET TO MULCH!

Covering pathways and the tops of beds underneath established crops with organic or inorganic mulch can drastically reduce weed pressure and time spent pulling weeds. See pages 94 and 130 for more information on choosing and using mulches.

CULTIVATING TOOLS (HOES)

Weeding plays a big role in increasing the productivity of a garden. Gardeners and farmers tend to have particular preferences about the type of cultivating tool they use. Humans have been designing cultivators for thousands of years, and you can find one in pretty much any shape imaginable. These are a few of our favorites.

WEEDING WILL ALWAYS BE A PART OF ORGANIC GARDEN MAINTENANCE. YOUR GOAL IS SIMPLY TO REDUCE THE OVERALL PRODUCTION OF WEEDS IN YOUR GARDEN.

Hula hoe. Also known as a wiggle hoe or stirrup hoe, this is probably the most commonly used manual cultivating tool in American gardens. The business end of the hula hoe is shaped just like a stirrup on a horse saddle and the blade will move slightly forward and back as you weed so that it can cut weeds both coming and going (it is sharp on both the front and back of the blade). Standard technique is to run the bottom part of the blade just under the surface of the soil and push/pull the handle back and forth to sever weeds. In situations where weeds have grown out of control, you may want to weed with the hoe and then go back with a rake to gather up all of the accumulated debris.

Collinear hoe. A newer cultivating tool that is very popular with small-scale professional growers, the collinear hoe has a sharp, narrow blade that is great for weeding in small spaces between rows and individual plants. It's specifically designed for small weeds, so functions best when used regularly.

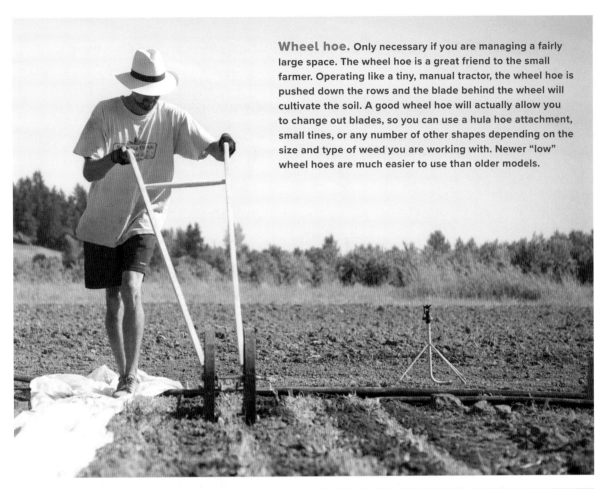

Wheel hoe. Only necessary if you are managing a fairly large space. The wheel hoe is a great friend to the small farmer. Operating like a tiny, manual tractor, the wheel hoe is pushed down the rows and the blade behind the wheel will cultivate the soil. A good wheel hoe will actually allow you to change out blades, so you can use a hula hoe attachment, small tines, or any number of other shapes depending on the size and type of weed you are working with. Newer "low" wheel hoes are much easier to use than older models.

Wire weeder. Very good for weeding between small and closely spaced crops, the wire weeder is so slight that it barely disturbs the surface of the soil. This is the type of hoe to use when you want to delicately remove very small weeds around sensitive plantings.

Wire hoe. Wire hoes (easy to confuse with the "wire weeder") come in a range of sizes for wider or more closely spaced crops. Their primary advantage is that they have no sharp edge, so the likelihood of damage to nearby crops is greatly reduced. Their lightweight heads are quite flexible and not appropriate for larger weeds.

PART
3

GET TO KNOW YOUR PLANTS

GROW MORE FOOD BY PLANTING THE RIGHT VARIETIES AT THE RIGHT TIME WITH THE BEST CARE.

To get the highest yields possible, it's important to use your time and resources efficiently. Your goal is to ensure every single plant brings its A game. If care is taken with every step of the growing process, you'll produce large quantities of high-quality crops with less investment of time and energy.

CHAPTER SEVEN

——

SELECTING SEEDS

Your seeds are the backbone and foundation of your annual garden system. To maximize yields, you'll need to understand the full spectrum of seeds available. You can then choose seeds that are appropriate for your climate, resistant to the diseases in your area, and bred for productivity.

Placing your seed order as early in the season as possible will ensure that your actual seed library matches your expected garden plan. Highly desirable varieties often sell out; buying your seeds in December or January will guarantee that you have the seeds you want on hand when the season starts.

Keeping a well-stocked seed library will allow you to take advantage of production opportunities on your own schedule throughout the season. A robust seed collection will provide opportunities to replant failed crops or squeeze in extra successions when weather conditions shift. Growing from seed can dramatically increase your opportunities for better yields.

sourcing seeds

We recommend that you order seeds directly from suppliers whenever possible. This method requires some foresight, since there is a necessary delay between the seed order and delivery of the seed; however, seeds purchased this way will be much less expensive, and you'll have more options when selecting varieties and quantities. However, it still helps to know which local stores carry high-quality seeds. Sometimes you'll run out of a particular seed midseason and may want to do a quick succession planting without waiting to place a seed order.

GOOD REGIONAL SEED COMPANIES OFTEN CARRY VARIETIES THAT ARE WELL SUITED TO YOUR CLIMATE.

CHOOSING A SUPPLIER

We encourage you to choose suppliers based on your location and gardening priorities. Some growers will choose to focus primarily on heirloom crops for their garden, while some will want to ensure that all of their seeds are organically produced. Other growers may value these characteristics, but be willing to adjust their expectations in order to increase chances of a high yield or resistance to a particular disease.

Good regional seed companies often carry varieties that are well suited to your climate. It's worth checking in with local suppliers to see what they recommend. Remember not to trust their advice blindly—some "local" seed companies operate with very little gardening knowledge and are happy to sell you a variety that will perform miserably in your climate. Make sure the seed suppliers you choose have a good reputation.

ORDER SEEDS EARLY

We recommend working through the catalogs and websites of your preferred seed vendors as early as October or November. This way, the season's successes and failures are still in the front of your mind. Hopefully you've documented these successes and failures in your records (see Chapter 2), but it's best to make decisions when your gardening challenges are fresh in your memory.

Try to place your seed order by the end of January, or earlier if possible. We place our orders in mid-December. The only drawback to a super-early submission is that you might find out about a new variety late in winter that you want to order. You can add on to your seed order through the season if this happens, but you might incur extra shipping fees or small-order fees. This is another good reason to talk with other local growers through the season to hear how crops are doing and get hot tips.

Many crops, like tomatoes, come in dozens or even hundreds of varieties.

selecting your varieties

↓ The moment you pick up a seed catalog or search a seed supplier's website, you'll realize the crops you're looking for come in seemingly endless varieties. This makes the process more exciting, but also more challenging. Some plants are bred or selected to produce the highest yields or the best flavor; others are bred for unique colors, adaptability to a particular climate, or disease resistance. Typically, a vegetable variety offered today has a blend of desirable qualities that makes it adaptable to various situations.

In catalogs and websites, seed suppliers will regale you with tales of success and wonder. While it may be that a variety does in fact grow well in some climates with a certain set of circumstances, not all of them will grow well in *your* garden. Be sure to read the catalog descriptions carefully; the best seed companies tell you what conditions a variety is best suited for. Sometimes, the name of the variety serves as an indicator for the conditions it likes best. For example, 'Winter Density' lettuce grows well in cool temperatures and 'Summerfest Komatsuna' has been bred to tolerate hot weather.

Local gardeners and farmers can be good sources for seed recommendations. Online websites and forums can also provide some useful insights. Internet research will take some sleuthing with very targeted searches, and it's important to keep in mind that there is no single online source that will tell you the best crops for your area. As you gain experience, you should rely more on your own observations than anything else to select the right varieties for your garden. Choosing crops and varieties that are adapted to your growing conditions will make your garden much more productive and successful.

Here are a couple of tactics for refining your variety selection from year to year.

+ **Grow several different varieties of each crop.** This way, you'll be able to compare them and determine which are most productive in your climate and in your garden. In addition, some varieties are best suited for early-season planting and some for late-season planting. One might be best for winter storage while another is ideal for fresh eating. Planting multiple varieties will also increase the genetic diversity in the garden. Some varieties are more disease, drought, heat, or cold resistant than others, so planting different types can act as a form of crop insurance.

+ **Ditch the duds.** Eliminating varieties that do not perform well in your garden is an important step in increasing your garden's productivity. Each season, we recommend that you experiment with a few new varieties, make sure to include some of your reliable favorites, and strike others from the list.

HYBRID OR OPEN-POLLINATED?

When deciding which varieties to choose, you may want to consider the crop's genetic history. Crops fall into two genetic categories: hybrid

and open-pollinated. Hybrid crops are the result of breeding two distinct genetic lines and then mating those lines to create an offspring with the best attributes of each parent. Open-pollinated crops are bred as a single genetic line over a longer period of time, slowly improving over generations. Many open-pollinated crops are considered "heirloom" varieties. This designation is vague, but indicates that the variety has been grown for a long time and has stable genetics. All heirlooms are open-pollinated, but not all open-pollinated crops are heirlooms. For a number of reasons, we feel that it's important to use a combination of hybrid and open-pollinated crops.

WHAT HYBRIDS OFFER

Many hybrid varieties are more resistant to pests and diseases, so they're well suited to organic production methods. Additionally, many hybrids grow more quickly than their heirloom cousins, so are well suited to challenging climates. Hybrids that have been bred for conditions that match yours can be much healthier and more productive than open-pollinated crops, leading to higher yields, lower mortality rates, and easier overall garden management.

For example, most open-pollinated peppers have a difficult time ripening in the cool nights of the Pacific Northwest. Hybrid varieties of peppers can be much more productive in this climate. Similarly, downy mildew can be a terrible problem for spinach growers in wet climates, but by selecting a downy mildew–resistant hybrid variety, it's possible to grow a delicious crop of spinach without using fungicides.

THE BENEFITS OF OPEN-POLLINATION

Open-pollinated varieties have a number of positive attributes working in their favor. In particular, heirloom open-pollinated crops are often thought to be better tasting than new varieties or hybrids; sometimes they really are. It's important to note, though, that heirloom varieties aren't

'Winterbor' kale has been bred to thrive in cold conditions.

inherently better tasting than hybrids—it's just that heirloom varieties were often selected for taste rather than other genetic attributes. In addition, some open-pollinated varieties have been selected for their excellent disease resistance. On the flip side, some varieties may be more difficult to grow, less tolerant of adverse growing conditions, and may produce lower yields. It's important to do your research.

If you're interested in seed saving, you'll need to choose open-pollinated varieties. Only open-pollinated crops will produce seed that is viable and that breeds true to itself. Breeding true means that the plants grown from your collected seed will produce a crop with similar traits (growth habit, disease resistance, taste, etc.) to the parent plant. Even with open-pollinated crops, to keep the traits you want, you'll need to learn about their genetics and how to separate flowering crops by time and distance in the garden. Seed saving is an art unto itself and beyond the scope of this book. If you're interested in this topic, we highly recommend picking up a few of the books listed in Resources (page 292).

Whatever their attributes or drawbacks, we feel it's vitally important to grow open-pollinated crops, simply for the sake of protecting the biodiversity of our food crops. The best way to protect and preserve these crops is to grow them, eat them, and save their seeds to replant. We also feel it's important to support responsible seed companies and the plant-breeding work they do, so we recommend that you grow a mixture of open-pollinated heirlooms and hybrid crops. Over time, you will continue to select new and different varieties based on your preferences and on the demands of your local conditions.

SELECTING VARIETIES FOR SEASONALITY

Seed descriptions will indicate if a variety was bred to tolerate specific weather conditions. This may be included in the variety name, as in 'Black Summer' pak choi. Alternatively, it may be nestled into the larger description of the crop; it might say "great summer variety" or "best when planted for spring production." A description may even call out a certain region of the country in which the variety does best, such as "great for northern regions" or "adapted for fall planting in the South."

All of these indications are crucial because they will tell you when and where to sow the crop for best performance. For example, you've probably noticed that there are hundreds of varieties of head lettuce. Lettuce is well known for its intolerance of hot weather. Summer heat makes it bitter and causes bolting. Because this is such an issue for growers, breeders have worked to develop varieties that perform better in midsummer conditions. The names and descriptions of the lettuce varieties will tell you whether the seeds should be sown early in the season, in the middle of the season, or late in the season.

If you want a consistent supply of head lettuce, you will need to use different varieties through the season, changing the type as the weather progresses. 'Winter Density' is a romaine type that grows best in cool temperatures and 'Jericho' is able to tolerate warmer summer weather, so you may want to plant 'Winter Density' in spring and fall, but sow 'Jericho' all summer long.

CHOOSING CROPS FOR CLIMATE

Certain crops are suited for particular climates. Crops that evolved from wild plants in tropical regions will always be more challenging to grow in cool climates. For example, tomatoes, peppers, and eggplant are heat-loving crops that will produce for incredibly long seasons in Georgia, but very short seasons in the Northeast and Pacific Northwest.

There are some crops like peanuts, sweet potatoes, watermelons, and okra that require so much heat that they are not even appropriate to plant in cool climates. Similarly, there are crops like spinach and broccoli that perform better and produce greater yields in cooler climates. You're likely to choose crops from all regions of the world, but be sure to match your planting plan and your variety selections to your local climate.

DISEASE RESISTANCE

Plant breeders have worked hard to create varieties that are more resistant to disease than their predecessors were. Indications of disease resistance are usually noted in a catalog's crop description. In fact, because these traits are so desirable, disease resistance is usually advertised prominently. Disease resistance is often written in notation, so "PM" will refer to powdery mildew and "CMV" will refer to cucumber mosaic virus.

Before you purchase a disease-resistant variety, however, you should know which diseases are prevalent in your region. Selecting a variety specifically for its resistance to CMV will not be helpful if this disease isn't a problem in your garden to begin with. It's not generally a problem to plant a variety that's resistant to a disease your garden doesn't experience, but sometimes varieties bred for disease resistance have other less desirable traits. For example, the crop may produce lower yields or smaller fruit than another variety would.

HARVEST WINDOW & STORAGE CAPACITY

Even closely related varieties will show differences in their ability to hold quality in the field and in the pantry. For crops that have the tendency to bolt (flower prematurely), look for varieties described as "slow to bolt." Take note of variety names and descriptions to ensure that if you intend to store potatoes through winter, you're planting a variety specifically indicated as a storage potato. Similarly, other root crops like carrots can have widely varying storage lives. Keep in mind what your particular goals are for the crop and make sure to order types that will meet that goal.

PLANT FOR TASTE

There's no doubt that your gardening experience will be much more satisfying and useful if you're producing crops that you (and the other benefactors of the garden) like to eat. Since catalog

CHOOSING THE RIGHT CARROT FOR YOUR GOALS

Carrot varieties can perform very differently in different soil types and have varying storage capabilities. If you want your carrots to yield well in heavy soil, you need to choose the right type. If you want to be eating carrots from your root cellar all winter, you need to choose the right type for that as well. Here are some basic differences between the major carrot types to guide you in your quest.

× **Danvers.** Mid-length (6–7 inches), conical shape with wide "shoulders" (the top part of the root). Usually best cooked. Widely adapted, produces best in sandy or well-balanced soil. Stores well.

× **Chantenay.** Short (5–6 inches), conical shape with a blunt tip and wider shoulders. Best type for clayey, heavy soils. (A farmer we knew in Pennsylvania swore he could not grow carrots due to his rocky clay soil, but after trying Chantenay types, they became a staple in his CSA program.) Many varieties are excellent for fresh eating and cooking. Stores well.

× **Imperator.** Long (up to 10 inches), gently tapering shape. This is the largest/highest yielding carrot type per square foot, but needs sandy or very well-balanced soil to perform well. Often more fibrous and less sweet than other types, so are best for cooking. Stores well.

× **Kuroda.** Short (5–6 inches), conical shape similar to Chantenay, but with less pronounced shoulders. This is another type that will grow in rocky/clayey soils that reject other carrot types. Commonly grown in Asia. Stores well.

× **Nantes.** Mid-length (6–7 inches), slender, cylindrical shape. This is our favorite type for flavor and texture when eating fresh, but it doesn't do well in heavy/rocky soil. Doesn't store as well as other types.

descriptions can only do so much to relate the flavors of crops, it will take some time to identify the varieties you like best. You can make quicker progress by asking for variety names at the farmers' market when purchasing vegetables and by talking with other gardeners in your area.

PLANT FOR PRODUCTIVITY

The top-line attribute for some crop varieties is actually their productivity. Names like 'Provider' beans or 'Sweet Million' tomatoes clearly signal that they are abundant producers. Look for detailed information on a variety's yield potential in the catalog description—some seed companies even provide charts showing relative yield between varieties of a given crop. You may find other high-yielding varieties by trial and error in your own garden, or by talking to other experienced local growers. Keeping good records will make sure you keep track of yield all-stars for future planting.

efficient seed ordering

↓ After you've chosen the specific varieties of crops you'd like to grow, you'll need to determine the quantities of seed to order.

FEWER MAY BE PLENTY

The quantity of seed you'll need will vary depending on the crop, on the number of plants you're planning on growing, and on whether you'll be sowing succession plantings. If you're growing a single row of tomato plants, for example, you can probably get away with ordering the smallest quantity available of any tomato variety (usually called a packet). However, if you hope to grow enough for your own garden *and* sell several hundred tomato transplants in spring, you'll obviously need to scale up.

Other long-season crops that need a lot of space, such as winter squash and tomatillos, might fall into a similar category; a small packet of seeds should be plenty for most growers. In addition, if you're trialing a new variety but don't know if it will perform well in your area or if you'll like the taste (say, you want to experiment with watermelon radishes), this is another good time to order a packet. If the crop performs well, it's easy to reorder for later in the season or for the following season.

Not all seeds store equally well. The seeds of some crops, like tomatoes and lettuce, maintain viability for 5 to 6 years or more, while alliums and parsnips have very limited viability and may keep only for a single year. Take this into consideration when ordering your seeds. You can purchase a 3-year supply of a tomato variety that you know you like, but you may have to order onion seeds every season. Keep in mind that ordering a larger quantity of seeds often decreases your per-seed cost, so it can be worthwhile to order a larger quantity if you know you can use the seeds over several years. See the Seed Life Span Chart (page 114) for more information.

The quantity of seed you purchase will depend on the crop's life span, how often you plan to succession plant, and the seed's storage viability. Shown here (left to right): 'Sugar Snap' peas, 'Early Wonder' beets, 'Borlotti' drying beans, 'Striped Armenian' cucumbers, and 'Brandywine' tomatoes.

ENOUGH FOR SEVERAL SOWINGS

If you're planning for multiple successions of short-season crops like lettuce or radishes, you'll likely need to order significant quantities of seed. Most good seed catalogs will have information to help you determine how much seed to order for these types of crops.

The catalog should indicate how many seeds to sow per row foot for direct-seeded crops, or how many are needed to produce transplants for X number of row feet. They also may tell you how many seeds are in an ounce or a pound. We find the Johnny's Selected Seeds catalog to be one of the best resources available for this type of information (see Resources, page 292).

Using information from the planting plan you created (see page 39), you can determine how many row feet or number of transplants for each crop you'll be growing. A bit of quick math will give you the number or quantity of seeds you need to purchase. Plan to order at least 20 percent more seeds than you calculate to compensate for low germination and other variables. Most seeds will remain viable for 2 to 3 years, so you can always use the extra in following seasons (see Storing Seeds for Increased Longevity, facing page).

Even with the best planning, you'll likely run out of or overorder certain seeds. Don't worry! With good record keeping, your seed-ordering skills will improve with every season. You can download the spreadsheet that we use for calculating seed quantities (see Resources, page 292).

SAMPLE SEED ORDER & INVENTORY TRACKER

CROP NAME	VARIETY	SUPPLIER	QUANTITY IN INVENTORY	QUANTITY TO ORDER	NOTES
Arugula	'Roquette'	Fedco	½ oz	1 lb	Generally reliable, still tastes good when it gets large.
Arugula	'Surrey'	Johnny's	0	¼ lb	I want to grow a lot more arugula this year and want to mix it up.
Beets	'Early Wonder'	Fedco	6 oz	0	Hopefully this will be enough seed for this season.
Broccoli	'Green Magic'	Fedco	0	Packet	New variety to try this year for summer.
Broccoli	'Marathon'	Johnny's	3 seeds	1 gram	Did really well last year; grow more this year (fall only).

storing seeds for increased longevity

The lifetime of a dormant seed can be anywhere from one season to 10 years and beyond. Most vegetable seeds have the potential to remain viable for at least 2 or 3 years, but this is hugely dependent on how they're stored. Seeds are often packaged in paper envelopes; they can be stored this way indefinitely, as long as they are kept dry and cool at all times.

Most seeds will keep longest if they're placed in a resealable plastic bag or airtight container and stored in the refrigerator or freezer. This is especially true for seeds that have a particularly short storage life. For example, if you'd like to keep allium seeds for more than one season, be sure to put them in the freezer. Keep in mind that every time you move a seed packet from the refrigerator to a warm room, moisture will condense on it, which could limit the storage life of the seed. In light of this, we recommend room-temperature storage for seeds that will be succession planted frequently throughout the year.

In fact, since fridge space is often at a premium anyway, you might store *all* of your seeds at room temperature in an area that remains consistently dark and dry. Good locations are a closet, cabinet, or clothes dresser in a room without south-facing windows. A garage or tool shed may not be a great location because of the wide variability in temperature and humidity through the year.

As seasons pass, use up your existing store of seed before opening a new package. If you notice germination rates declining, simply plant more seeds to compensate. If plants fail to come up at all, compost the seed and remember next time to order less or to use it more quickly.

A case designed for photo storage makes a perfect seed organization box. Larger bags of seed are kept in a larger box, and those used for frequent direct-seeding successions are kept in spice shakers.

SEED LIFE SPAN CHART

ANNUAL VEGETABLES AND HERBS	LATIN NAME	APPROXIMATE SEED STORAGE LIFE SPAN (YEARS)
Arugula	*Eruca vesicaria*	3
Basil	*Ocimum basilicum*	3
Beans, edible soy (edamame)	*Glycine max*	3
Beans, fava (broad)	*Vicia faba*	3
Beans, lima	*Phaseolus lunatus*	3
Beans, shell (fresh or dried)	*Phaseolus vulgaris*	3
Beans, snap	*Phaseolus vulgaris*	3
Beets	*Beta vulgaris*	4
Brocccoli raab	*Brassica rapa*	4
Broccoli	*Brassica oleracea*	3
Brussels sprouts	*Brassica oleracea*	4
Cabbage	*Brassica oleracea*	4
Cabbage, Chinese	*Brassica rapa*	4
Carrots	*Daucus carota*	3
Cauliflower	*Brassica oleracea*	4
Celeriac	*Apium graveolens*	5
Celery	*Apium graveolens*	5
Chard, Swiss	*Beta vulgaris cicla*	4
Cilantro	*Coriandrum sativum*	2
Collards	*Brassica oleracea*	5
Corn, sweet	*Zea mays*	2
Cucumbers	*Cucumis sativus*	5
Dill	*Anethum graveolens*	3
Eggplant	*Solanum melongena*	4
Endive	*Cichorium endivia*	5
Fennel, bulbing	*Foeniculum vulgare*	4
Garlic	*Allium sativum*	1
Kale	*Brassica oleracea*	4
Kohlrabi	*Brassica oleracea*	4
Leeks	*Allium porrum*	1

ANNUAL VEGETABLES AND HERBS	LATIN NAME	APPROXIMATE SEED STORAGE LIFE SPAN (YEARS)
Lettuce, baby mix	*Lactuca sativa*	5
Lettuce, head	*Lactuca sativa*	5
Mâche	*Valerianella locusta*	5
Melons	*Cucumis melo*	5
Mustard greens	*Brassica* species	4
Okra	*Abelmoschus esculentus*	2
Onions, bulb	*Allium cepa*	1
Pak Choi	*Brassica rapa*	4
Parsley	*Petroselinum crispum*	1
Parsnips	*Pastinaca sativa*	1
Peanut	*Arachis hypogaea*	1
Peas, shelling	*Pisum sativum*	3
Peas, snap	*Pisum sativum*	3
Peppers, hot	*Capsicum annuum*	2
Peppers, sweet	*Capsicum annuum*	2
Potatoes	*Solanum tuberosum*	1
Radicchio	*Cichorium intybus*	4
Radishes	*Rhaphanus sativus*	5
Rutabagas	*Brassica napobrassica*	5
Scallions	*Allium wakegi*	1
Spinach	*Spinacia oleracea*	2
Squash, gourds	*Cucurbita* species	4
Squash, pumpkins	*Cucurbita pepo, Cucurbita maxima*	4
Squash, summer	*Cucurbita pepo*	4
Squash, winter	*Cucurbita* species	4
Tomatillos	*Physalis ixocarpa*	4
Tomatoes	*Lycopersicon esculentum*	4
Turnips	*Brassica rapa*	5
Watermelon	*Citrullus lanatus*	4

improve yields with seed treatments

↓ Some seed will benefit from a special treatment before or at planting time to ensure their success. These treatments may or may not be necessary depending on the crop and the treatment, but it's good to know how they work so you can include them in your tool kit of seeding techniques.

BACTERIAL OR FUNGAL INOCULATION

In the context of gardening, "inoculation" refers to introducing a microbiological agent to your growing system for the benefit of the crop. Common garden inoculants are bacteria or fungi that are intentionally added to the soil or seed of a crop to promote a healthier plant. By adding certain microbes directly to the soil, you can help create an especially robust soil ecosystem in your garden.

Rhizobium. This popular inoculant is a bacterium that is added to the seeds of legume crops at planting time. It comes in many different strains and is so widely used that it's often sold simply as "garden inoculant." Luckily, the most common garden legumes—peas and beans—are supported by the same strain. Additional strains exist for soybeans, peanuts, and leguminous cover crops. The bacteria form a symbiotic relationship with the root nodules of the crops, which allows the plants to absorb nitrogen from the air, use it for growth, and deposit it into the soil.

This process is called nitrogen fixation and is a unique trait of legume crops. Even though the air we breathe is 78 percent nitrogen ($N2$), plants

Legume crops may be healthier and more productive if inoculated with rhizobium bacteria.

can't use it in this form. It must be converted into ammonium nitrogen (NH4+) or nitrate nitrogen (NO3-) before plants can absorb it and use it for growth. Rhizobium bacteria allow the legumes to convert N2 from the air into NH4. This is helpful for the legume crop, but also leaves behind a store of NH4 in the soil for the next plants in the garden to use.

Inoculate the seeds just prior to planting by making a slurry. Take the seeds you need for that round of planting, pour them into a plastic or glass container, sprinkle inoculant onto the seeds, add a few drops of water, and stir until the powder sticks to the seeds and coats them lightly. The packaging will include an application rate, but the exact proportion of powder to seed is not crucial. A little bit goes a long way, and there are no adverse effects if you use too much.

Mycorrhizal fungi. Another, less commonly used but effective inoculant is made with mycorrhizal fungi. Mycorrhizae are soil-dwelling fungi that have been shown to help a wide range of plants build more robust and healthy root systems. Similar to the bacterial relationship described above, these fungi symbiotically colonize plants' root systems. In exchange for carbohydrates from the plant, the fungi expand the plant's underground network, helping the root system absorb water and nutrients from the farther reaches of the soil. Mycorrhizal fungi are naturally present in the soil, especially in a well-managed organic garden. However, the addition of extra fungi can boost plant growth, especially in new or poor garden soil that has not developed a robust soil ecosystem.

These fungi can be purchased in liquid, powder, or granular form. Fungi that are appropriate for use in the garden will be referred to as endomycorrhiza (in contrast to ectomycorrhiza, which live on the roots of woody plants). Powdered or granular forms are easy to apply when transplanting crops by simply adding the material to the planting hole as you would with a granular fertilizer. Liquid forms are easy to apply

Soaking seeds like parsnips can help speed up and improve germination rates.

to direct-seeded crops when disturbing the root system is undesirable or impossible.

OTHER SEED TREATMENTS

Certain seeds benefit from some kind of physical treatment. The two most common kinds of treatment for annual vegetable crops are scarification and soaking.

Scarification. This refers to the abrasion of the seed coat in order to prepare the seed for germination. It's applicable to seeds with particularly thick seed coats. Common techniques include rubbing with sandpaper and cracking with a hammer. Although not necessary for edible crops, it can speed up germination times for large seeds like beans and squash.

Soaking. Some seeds will sprout more quickly if soaked in water prior to seeding. This technique is particularly effective for cucurbits, peas, and beans. It can also be effective for slow-germinating crops like carrots, parsnips, and parsley. Soak seeds in lukewarm or cool water for up to 12 hours. Some growers add liquid kelp to their seed-soaking water at ¼ teaspoon per quart of water, which is reported to speed up the germination process even further.

CHAPTER EIGHT

TRANSPLANTING & DIRECT SEEDING IN THE GARDEN

Choosing whether to transplant or direct seed a crop can be crucial to success in your garden. Although transplanting requires more work at the outset, it can give your crops a leg up on some pests and an early start in areas with short growing seasons. It also reduces time spent thinning crops later on and avoids wasting space on unhealthy specimens. Direct seeding—also called direct sowing—gives you an overall quicker seed-to-harvest timeline. It's essential for many root crops, and for crops that don't yield well when transplanted. Direct seeding also allows you to plant large areas of the garden quickly and can minimize the overall investment in easy-to-care-for crops like beans, peas, and salad greens.

The particular method (or combination of methods) you choose for each crop will depend on the crop's growth habits, your climate, and your preferred management practices. Both methods require careful bed preparation, weather monitoring, and attention to detail.

Brassicas can take more quickly to the soil if their roots are disturbed before planting from a plastic container.

careful transplanting = productive plants

↓ Whether you grow your transplants or purchase them, you'll want to select the healthiest-looking specimens to go in the garden. It's not a good use of time, space, or resources to try and coddle along crops that are stressed from the outset. Be ruthless in your assessments and cull anemic, wilted, or otherwise off-looking plants before they ever go into your beds.

Even with the healthiest seedlings, the care you take when bedding them in can dramatically impact their success and productivity. If a plant is moved under stressful conditions, it may undergo "transplant shock," and be unable to successfully send new roots into the soil. Transplant shock can lead to wilting and stunted growth, and can reduce the overall yield of the plant.

PREPARE THE BED

Pull back the covers, climb in, and lie down for a nice nap. Oh wait, wrong kind of bed. Before bringing your plants to the site, prepare the garden bed thoroughly. Spread fertilizer (either broadcast or add to individual planting holes), check the irrigation system for leaks, and set lines in the appropriate location to help guide planting.

You can either predig your holes for transplants at the desired spacing, or remove the

PLANTING AT DIFFERENT DEPTHS

In most cases, the root mass of the transplant should be buried just below the surface of the soil in the bed. If the top of the root mass is not covered with soil, exposed root tips will be damaged and the entire root mass may dry out through evaporation. However, even though the roots should be completely covered, the stem of the transplant should be left exposed to minimize the chance of stem rot. There are several notable exceptions to this planting technique.

Brassicas. You can bury brassica stems without worrying about stem rot. In fact, young brassicas are notorious for their fragile stems, and burying the base of the plant up to the level of the first true leaves will actually help stabilize the plant in the soil.

Planting alliums

Trenching a tomato

Alliums. The stem and leaf portions of most alliums are planted deeply into the soil when transplanted. Bury leek stems 6 to 9 inches deep into the soil to help blanch them; this results in a larger and more tender harvest. The same technique can be applied to bulbing onions and scallions. Keep the top 3 to 4 inches of the plant above ground, but the majority of the young plant can be buried.

Tomatoes. A tomato plant will set roots from any portion of its stem that comes into contact with the soil. To take advantage of this vigorous rooting mechanism, remove the lowest sets of leaves from the transplant and bury the stem in the soil. You can bury the plant up to the top two sets of leaves. Some growers prefer to dig a shallow trench and bury the stem sideways in the soil, turning up the tip of the plant above the soil. This keeps more of the root mass at a shallow depth where the soil is warmer, but also carries the risk of breakage during planting time, since the stems are quite fragile.

Potatoes. Potatoes are planted from "seed potatoes." Seed potatoes are just healthy, disease-free individual potatoes from last year's harvest that are planted to grow a new crop. A single seed potato can produce up to 10 pounds of new tubers in a season. Plant small seed potatoes whole and cut larger ones into pieces to extend the planting stock. If you cut your seed potatoes, aim for a weight of 1 ounce per cut piece. Each piece must have a healthy looking eye on it; this is where the new potato vine emerges from.

Trenching potatoes

Burying brassica stems

Potatoes can be planted at virtually any depth, but deeper is typically better. To plant potatoes, don't initially cover the seed potato with any more than 4 inches of soil. As the vine grows, you can continue to add soil over the top of the plant as long as you leave the growing tip exposed to the sunlight. The more of the vine that is buried, the more tubers the plant will produce.

One way to plant potatoes is to dig a trench in the bed and leave the excavated soil in mounds alongside the trench. Bury the seed potato under 4 inches of soil and continue to add soil from the mound as the vine grows, until it reaches ground level and the soil is flat. Make sure the eye points upward when you place the piece in the planting hole or trench. Alternatively, you can plant the potato shallowly in a container and add soil as the plant grows. Anything that contains soil can be used, including straw bales, a trash can with drainage holes, or burlap sacks.

Transplanting works best when plants are watered well before removing them from their containers and planted quickly into well-prepared soil, then watered in immediately.

plants from their container and set them out across the bed at the desired spacing in preparation for transplanting. Once you have some experience, you can eyeball the spacing between plants pretty accurately. When starting out or getting accustomed to new spacing, measure between holes with your hand trowel or a ruler.

WATCH THE WEATHER

Ideal transplanting conditions are overcast skies, relatively cool temperatures, and even a slight rain. Weather conditions like this will minimize transpiration (water loss from plant leaves) and help ensure soil moisture levels remain high enough to reduce damage to root systems and encourage quick new root growth. These conditions will reduce overall transplant shock and allow plants to transition back to healthy growth rates as quickly as possible. If you have a lot of transplanting to do and the weather is hot and sunny, consider waiting a day or two if cool weather is in the forecast. If hot, dry weather is unavoidable, thoroughly water the garden soil and the starts before beginning transplanting. Try to do your transplanting early in the morning or in the evening to reduce heat stress.

HANDLING TRANSPLANTS GENTLY

Different crops tolerate a wide range of transplant conditions. The most tender plants tend to be cucurbits (squash, cucumbers, pumpkins) and legumes (beans, peas). When planting these sensitive crops, try to minimize disturbance to the roots as much as possible. Predigging transplanting holes is ideal. Very carefully remove each plant from its container, gently place it into the prepared hole, cover the roots with soil, gently compress the soil to help stabilize the plant, and immediately water it.

Brassicas, lettuce, and nightshades are much tougher. Some growers even recommend intentionally disturbing the roots of brassicas at transplanting time to encourage the development of new roots. To prepare a brassica transplant, tear the outer roots gently with your hands so they no longer hold the shape of the container they were grown in.

WATER WELL

It's important to water plants as quickly as possible after transplanting to reduce the amount of time their roots are exposed to air. Watering after the plant is in place will settle the soil around the root tips and protect them. Consider watering in the plants with a low dose (typically half of a regular dose) of liquid high-phosphorus organic fertilizer. Use a 2:4:2 or something similar, or see DIY Organic Fertilizers (page 139).

After transplanting and watering the plants, go around the bed and finger-test the soil for moisture near a few different plants. To finger-test, simply put your pointer finger down into the soil up to the knuckle. The soil should be evenly moist all the way to this depth. If your finger encounters dry space, continue watering the transplants until the soil is evenly moist on every finger test.

These beets germinated well along a drip irrigation line.

direct seeding

Direct seeding—planting the seeds of a crop directly into your garden beds—is a very efficient way to get a crop started. It requires much less time up front, and an entire bed of crops can be planted in just a few minutes. However, establishing and maintaining these plantings can require more work later on.

Making an appropriate choice between direct seeding and transplanting can greatly affect your yields. This choice is affected by many variables, including the space you have available in your garden (and nursery, if you have one), your general climate and microclimate, the conditions in your garden, and the crops you're growing.

Many professional growers will use both methods to plant the same crop at different times of the season. For example, in areas with cool spring weather, a grower might transplant their first planting of summer squash to ensure a solid stand and good early growth when the soil is cold, and then direct seed the second and third plantings when the soil has warmed enough for good germination. Consider the following benefits and drawbacks of direct seeding when making your choice.

BENEFITS OF DIRECT SEEDING

+ You don't need to manage plants in a propagation area, reducing the time, materials, cost, and attention needed to get the crop under way.

+ You can rely on rain to keep seeds moist during germination.

+ With sufficient rain or an irrigation system, crops need very little care during their early weeks of growing.

DRAWBACKS OF DIRECT SEEDING

+ Increased weed pressure when the plants are young and most difficult to weed around.

+ Attacks from insects when the plants are young and likely to be severely affected or killed by such damage.

+ The need to thin the plants to achieve desired spacing.

TRANSPLANTING PEAS & BEANS

Thoughtful planting strategies can seriously increase a garden's productivity. Peas and beans offer an interesting case study. Tradition dictates that these crops should be direct seeded into the garden. Since the seeds are large, are planted deeply enough that desiccation is unlikely, and have relatively high germination rates, in many situations direct seeding is the easiest way to manage plantings. However, since the big seeds and tender shoots are a favorite target of all sorts of garden pests, some growers prefer to transplant these crops, reducing the likelihood of a replanting. Transplanting will also allow for more succession plantings each season. Instead of clearing a spring crop and direct seeding a row of bush beans to replace it, you can transplant 2-week-old beans into the bed, saving valuable prime summer growing days.

+ The possibility of low germination rates due to the whims of nature (abnormally cold days, lack of rain, etc.).

+ Slower germination rates in cool soil conditions.

WHICH CROPS TO DIRECT SEED

Certain crops grow best when direct seeded. While these are not hard-and-fast rules, we find the following crops easiest to manage and most likely to succeed when sown directly into the garden bed.

Root crops. Typically, root crops—including beets, carrots, parsnips, radishes, and turnips—are best when direct seeded into the soil. These crops have been bred to grow substantial roots (your food source) at the expense of robust aboveground growth. Because the plant is root focused, pulling up and transplanting these crops can be stressful and lead to delayed or stunted growth, reduction in harvest, or death.

Baby greens. When growing *individual plants* for greens (head lettuce, frisée), it's helpful to grow them from transplants. However, when growing closely spaced densely seeded crops like baby lettuce, baby arugula, braising mix, spring mix, and most other loose-leaf cutting greens, they'll produce best and be easiest to manage when direct seeded into the garden.

Cucurbits. Due to their sensitive root systems, cucurbits (squash, cucumbers, pumpkins, melons) grow best when direct seeded. Assuming weather conditions are suitable, direct seeding these crops will eliminate the possibility of transplant shock and stunted growth. However, in many climates, the growing season is not long enough for direct-seeded cucurbits to reach maturity. In these situations, take extra care when transplanting these crops into the garden. Consider growing them in soil blocks or peat pots to help reduce handling during transplant. But don't be afraid to grow these crops in your preferred transplanting containers; good technique will accommodate for their sensitivity.

TIPS FOR SUCCESSFUL DIRECT SEEDING

Direct seeding crops is a fine art. By taking the time to properly prepare for and manage these plantings, you dramatically increase germination rates, improve plant health, and save yourself tons of time and effort later in the season.

Create a fine seedbed. Create an environment that will make the seed's life easy from day one. As a seed germinates and begins to push roots down and send up a shoot toward the sky, large soil particles, chunks of wood, or small rocks can create physical barriers to root and stem development. So, before you sow, smooth and flatten the surface of your garden bed and remove large particles or break them up into small pieces. Also, if your soil is particularly clumpy or dense, you can sprinkle germination mix or a screened soil mix over the top of a newly seeded crop.

Create a stale seedbed. Prepare your seedbed several weeks before you plan to direct seed your crop. Loosening the soil and irrigating the bed ahead of planting will encourage existing weed seeds to germinate. This will give you the opportunity to clear these weeds quickly and easily without worrying about disturbing your young crops. Prepare the bed, wait 2 weeks, cultivate the bed shallowly so as not to bring new weed seeds to the surface, direct seed your crop, and enjoy lower weed pressure. Tarping or flame weeding also work well for creating a stale seedbed (see pages 98–99). This technique is especially useful with slow-germinating crops like carrots or parsnips.

(see pages 98–99)

SOIL THERMOMETER

This is an inexpensive tool that can really help you keep an eye on real-time conditions in case you need to adjust planting schedules or monitor growth rates through the season. Different crops require different soil temperatures for germination; using a thermometer will allow you to adapt your seeding schedule based on actual measurements, rather than just calendar dates.

Sow at the optimum soil temperature for each crop. All seeds have an ideal soil temperature range for good germination. Soil that is too cold may rot seeds before they have the chance to sprout. Soil that is too warm may kill seeds that are sensitive to higher temperatures. Your seed packet and seed catalog should indicate the ideal germination temperature for your crop. Keep in mind that using a soil thermometer will give you a more accurate reading of how warm the soil is than simply testing the soil with your finger will.

Many root crops, like carrots, perform best when direct seeded into the garden.

WHEN TO BROADCAST SEED

Broadcasting is an alternative seeding technique for closely spaced crops. You sprinkle seeds over a given square footage of a bed, and then rake or mix them gently into the soil. This is a useful technique for planting baby salad greens because you can utilize the entire surface area of a bed for the crop. It's also useful for sowing small-seeded cover crops. We don't recommend this technique for other vegetable crops because it's difficult to manage seeding depth and makes weeding very difficult.

Irrigate your bed on planting day. Unless your soil is already moist on seeding day, plan to irrigate the area prior to seeding. Moistening the soil will help it settle slightly, and this will ensure that seeds are not washed out of place when you water them after sowing. Water the bed, sow your seeds, cover with soil, and water the area again.

Monitor soil moisture as seeds germinate. Because most seeds are sown very shallowly in the soil, they are at risk of drying out during the germination process. As the sun hits your beds, the top ¼ inch of soil (where most seeds are located) can dry out very quickly. Even if the majority of the soil remains moist, a dry upper crust can wreak havoc on germination rates. Unless you are receiving regular and adequate rainfall, plan to water newly seeded crops with sprinklers or by hand on a regular schedule until they have fully germinated.

Date your planting. Noting the planting date on the plant tag and in your records will allow you to keep an eye on the rate of germination. If a crop is expected to pop up in 5 to 7 days but no growth is seen in 2 weeks, you can reseed the crop right away without losing too many growing days. If you sow a crop and fail to monitor its days to germination, it will be difficult to know if the seeds are still viable or the area should be replanted.

Plant an appropriate quantity of seeds. There's a fine balance between overseeding, which creates crowding and difficult thinning, and underseeding, which results in sparse plantings that need to be resown or that fail entirely. Even though a seed packet might indicate that a crop has a very high germination rate, keep in mind that this rate is applicable only to ideal conditions. In an outdoor real-time application, germination rates will be much lower. The thickness and spacing will be different for every crop, but plan to sow at least two to three times as many plants as you want to harvest.

Sow in an orderly arrangement. Most crops are easiest to direct seed in rows or in other geometric arrangements. Keeping rows straight makes cultivating between and around crops much easier. Sowing directly along drip lines will also increase germination rates and make sure all parts of the planting receive adequate water throughout the season.

For closely spaced crops, try making a furrow at the desired seeding depth of the crop you're sowing, and sprinkle seeds at the desired rate in it. You can then firm soil over the furrow and water everything in. For example, carrots might be planted in ¼-inch-deep furrows at two to three seeds per inch, with furrows spaced 6 inches apart for cultivating.

For widely spaced crops like pumpkins or summer squash, place seeds directly in the ground at the desired spacing, and cover with soil to the desired depth. For example, you might place three cucumber seeds 1 inch deep in the ground every 2 feet in a bed.

Protect seeds as needed. Seeds and newly sprouted plants are particularly sensitive to pest damage. A single bite from an insect on a day-old plant may kill it. Similarly, birds have been known to swoop in and eat up a planting of direct-seeded crops before they've even had the chance to germinate. If you are concerned about the intrusion of a certain pest, take preventative actions. Depending on local pests and the crop, appropriate actions may include adding beneficial nematodes, covering seeds with floating row cover or bird netting, adding organic slug bait, applying diatomaceous earth, or putting out rodent traps.

THIN YOUR DIRECT-SEEDED PLANTINGS

In the home vegetable garden, where space is always at a premium, overseeding crops and thinning—culling plants to allow adequate room for each maturing crop—is often the most efficient way to use bed space and increase overall yield from the garden. Of course, this requires that you actually follow through on thinning the crops.

Thinning is a relatively time-intensive activity, though also very straightforward. Begin at one end of the row or planting and work your way toward the other end by removing the extra plants. If you've accidentally overseeded, you may end up removing dozens or even hundreds of plants. Keep in mind that, depending on the crop, you may be able to eat the thinning. So instead of lamenting the crops that could have been, be thankful for an especially early harvest.

If the sowing is particularly thick, or if you want to make sure you have enough viable plants after thinning, consider thinning in two passes. Once the crops have germinated and are tall enough to be easily distinguished and separated, thin the row to half of the eventual mature spacing. Two weeks later, make a second pass to thin to the final spacing. For example: If planting beets, the final spacing between plants should be 4 inches. On the first thinning, remove enough plants so each remaining beet is spaced 2 inches apart. On the second thinning, remove plants to achieve the final 4-inch spacing.

When direct seeded, beets often need thinning. The thinnings can be eaten as an early harvest.

INVEST IN A SEEDER

Of course, on a large scale, thinning can be impractical. If you are growing hundreds of row feet of direct-seeded crops, you may want to invest in a seeder. Seeders designed for small farms can be huge time-savers. A good seeder will have interchangeable rollers or disks that allow for precise spacing of any vegetable crop. EarthWay and Jang are reliable brands that make seeders adaptable for the large home farm. When used properly, they can virtually eliminate the need for thinning.

DIRECT SEED OR TRANSPLANT?

For crops that can either be direct seeded or started indoors, we often prefer to start them indoors if nursery space is not a limiting factor. We find that plants are more successful when started in a controlled environment, pampered for several weeks, and then placed into the garden at the ideal spacing required for maximum production. Also, in areas with short growing seasons, many crops (like tomatoes) must be transplanted to ensure they have time to grow to maturity before cold weather halts their production. Any crop that has a life cycle longer (or nearly as long) as your growing season should be started indoors or purchased as a transplant.

TP: Transplant   DS: Direct seed   E: Either
B: Both; transplant early successions and direct seed later successions
We suggest starting plants listed as "TP, E, B" indoors and transplanting to the garden, especially early in the season. This does not mean the crop cannot also be direct seeded, however. If you prefer, these crops can be direct seeded or you can do plantings each way.

ANNUAL VEGETABLES & HERBS	PLANTING TECHNIQUE
Arugula	DS
Basil	B
Beans, edible soy (edamame)	E
Beans, fava (broad)	DS
Beans, lima	E
Beans, shell (fresh or dried)	E, B
Beans, snap	E, B
Beets	DS
Broccoli	TP, E, B
Broccoli raab	TP, E, B
Brussels sprouts	TP, E, B
Cabbage	TP, E, B

ANNUAL VEGETABLES & HERBS	PLANTING TECHNIQUE
Cabbage, Chinese	TP, E, B
Carrots	DS
Cauliflower	TP, E, B
Celeriac	TP
Celery	TP
Chard, Swiss	TP, E, B
Cilantro	DS
Collards	TP, E, B
Corn, sweet	E
Cucumbers	TP, E, B
Dill	TP, E, B
Eggplant	TP
Endive	TP, E, B

ANNUAL VEGETABLES & HERBS	PLANTING TECHNIQUE
Fennel, bulbing	TP, E, B
Garlic	DS
Kale	TP, E, B
Kohlrabi	TP, E, B
Leeks	TP, E, B
Lettuce, baby mix	DS
Lettuce, head	TP, E, B
Mâche	TP, E, B
Melon, cantaloupe, honeydew	TP, E, B
Mustard greens	DS
Okra	TP, E, B
Onions, bulb	TP
Pak choi	TP, E, B
Parsley	TP, E, B
Parsnips	DS
Peanut	TP, E, B
Peas, shelling	TP, E, B
Peas, snap	TP, E, B
Peppers, hot	TP

ANNUAL VEGETABLES & HERBS	PLANTING TECHNIQUE
Peppers, sweet	TP
Potatoes	DS
Radicchio	TP, E, B
Radishes	DS
Rutabagas	DS
Scallions	TP, E, B
Spinach	TP, E, B
Squash, gourds	TP, E, B
Squash, pumpkins	TP, E, B
Squash, summer	TP, E, B
Squash, winter	TP, E, B
Sweet potatoes	Slips
Tomatillos	TP, E, B
Tomatoes	TP
Turnips	DS
Watermelon	TP, E, B

mulch at planting time

Mulching is a great way to minimize weeding and moisture evaporation from the soil. In fact, mulching may reduce irrigation needs by 50 to 80 percent, depending on your climate. It'll also prevent erosion during periods of heavy rainfall, help regulate soil temperatures, and create a better environment for beneficial soil organisms. Mulching in the garden generally works best with transplanted or widely spaced crops.

For ease of application, consider spreading the mulch first, then making holes in it for setting transplants or direct seeding. If you use drip irrigation, lay the lines underneath the mulch. With transplants, you can water the crop in well and then immediately pull the mulch back around the plant stem. When direct seeding, wait until the plant germinates and grows a bit before pulling the mulch around the stem.

Here are a few cautionary notes on mulch.

+ Mulch increases moisture levels and decreases airflow around the base of plants, so it can make fungal problems worse. If your garden has had problems with damping-off in the past, wait to spread mulch until later in the season when the plants are well established.

+ In-season mulching for closely spaced direct-seeded crops like lettuce mix and carrots is not recommended.

+ Mulch provides a great habitat for slugs and other dark- and damp-loving insects, so it may not be the best technique if you have serious problems with these critters.

+ Most mulching materials contain a lot of carbon, so it's best to not mix them into the soil unless they're very well broken down. Wait to incorporate mulches until the end of the growing season or winter. Consider raking them off and composting them instead. If you've used mulch during the growing season and it's still in good condition, you can leave it in place for winter protection.

+ Be careful when using bark mulch, sawdust, and wood chips on beds. They make a great mulch around perennial crops and trees, and can be effective as mulch for annual vegetables; however, it's critical for annual crops that they not be mixed into the soil—their high carbon content means that as they decompose, they can tie up nitrogen. They can also be difficult to rake off beds at planting time, especially if you're preparing to direct seed a small-seeded crop like carrots or lettuce.

AVOID TYING UP NITROGEN

Fine, highly carbonaceous material such as wood chips, bark mulch, and sawdust can cause problems if they are mixed into vegetable garden soil. While these materials will eventually decompose and release beneficial nutrients, they can stunt the growth of your crops in the short term because they tie up nitrogen.

In order to grow and reproduce, soil organisms need to consume organic materials. When there is enough nitrogen-rich material, the microbes burn nitrogen compounds for energy and release inorganic nitrogen compounds like ammonium and nitrate. This is called mineralization, and it produces forms of nitrogen that are available to plant roots. Immobilization (nitrogen tie-up) takes place when you incorporate high-carbon/low-nitrogen materials into the soil. In this instance, the microbes take ammonium and nitrate from the soil in order to meet their nitrogen requirements, and temporarily render it unavailable to your crops.

MULCHES FOR THE GROWING SEASON

- × **Compost.** Mulching with compost midseason supplies nutrients to your crops as precipitation passes through it.

- × **Grass clippings.** Don't let them contact plant stems directly as they can get hot during decomposition.

- × **Shredded hardwood leaves.** Can be left on the soil over winter and turned into the soil in spring to add organic matter and nutrients.

- × **Plastic mulch.** Helps warm soil and prevent weeds (though it can be a chore to remove).

- × **Landscape fabric.** Works well for weed prevention and can be reused from year to year. Try laying it down first, then cutting holes at appropriate spacing for transplants.

- × **Rolled burlap or burlap sacks.** An effective mulch, but can be difficult to lay in between plants. Can be reused from year to year.

- × **Straw.** Use a thick layer (4 to 6 inches). If using straw as a winter mulch, rake it off in the spring and compost it or reapply it when crops mature. Turning it into the soil could tie up nitrogen. Be sure to confirm that the straw you're buying was not sprayed with chemical herbicides. These can persist in the soil for years, so adding treated straw to your garden could lead to low germination rates and crop health issues for a long time into the future. It is also essential that the straw not have seed heads. Adding straw with seeds can create a much more intense weed problem than the one you were trying to solve.

Landscape fabric can be an effective mulch for superlong-season crops like garlic.

Cucamelons grow up on a permanent metal mesh trellis.

Peppers can be trellised with bamboo using the stake and weave method.

support climbing plants

Vining plants will take up less space in the garden, be less susceptible to pest and disease issues, be more productive, and be easier to harvest when trellised. Some crops have tendrils that help them climb up your support structures, and others will depend entirely on you to help keep them off the ground. With a few exceptions, all vining crops should be trellised whenever possible. A little bit of extra help can even keep shorter crops from falling over if they have strange growth forms or heavy loads of fruit.

Planting time is the best time to set up the trellises and other support systems you'll be using. This will prevent disturbing your plants later on and will make directing plants onto the trellis much easier. Plant stems can be floppy and fragile, so attempting to redirect them forcefully while they're growing is asking for trouble.

Often, plants will need some assistance locating and climbing a trellis. Keep an eye on your crops as they grow and help them locate the structure and guide them as much as necessary. Don't hesitate to redirect an errant branch, or to tie them loosely to the structure with twine if needed.

In general, trellises:

+ provide good air circulation, which helps to reduce pest and disease pressure.

+ keep the fruit off the ground, reducing damage from ground-dwelling insects and animals such as mice and rats.

These peas are trellised along a fence and secured with twine every 3 to 4 inches.

Cucumbers can be easily trellised on twine. Even vining crops may benefit from additional support like vine clips.

+ allow for better sun exposure on more of the plant's leaves.

+ give branches support, preventing them from breaking due to heavy fruits or heavy winds.

+ help utilize space more efficiently; growing vertically allows you to grow more food in the same amount of space.

You'll also benefit from the improved appearance of your garden—tall, well-managed crops add a unique structure and beauty to vegetable beds.

TYPES OF SUPPORTS

Some trellises can be easily built with common household materials (twine, wire, etc.), though others may require prefabricated structures (tomato cages, lattice, etc.). You may even want to purchase building materials and make your own; a beautiful, functional trellis can be made from materials found at any hardware store or commercial nursery.

TWINE

By attaching a string of twine to an overhead structure, you can create a simple trellis for tomatoes, pole beans, cucumbers, or any other vining crop. Tie a piece of twine to a sturdy overhead structure or use a time-saving device like a Tomahook to secure the upper part of the string. Loop the lower end loosely around the base of your plant. We find that this is a very cost-effective solution when trellising large numbers of vining crops. As a bonus, some twines can be cut down and composted along with the plants at the end of the season, reducing workload and off-season trellis storage needs.

The disadvantage is that crops without tendrils often need ongoing assistance to make sure they stay on course and remain connected to or wrapped around the twine. If you're growing a nonvining variety of a crop like tomatoes, make it a weekly practice to wrap the new growth around the twine. Frequent management will ensure you're never trying to force an older, more fragile

Tomatoes can be trellised with individual rebar posts.

THE BEST TIME TO SET UP SUPPORT SYSTEMS IS WHEN YOU'RE PLANTING.

the vines once they are growing and attached to the structure, so they need very little midseason maintenance. The downsides are that finding and harvesting ripe fruits can be more of a challenge in a cage system and the fact that cages take up a ton of storage space when not in use.

SINGLE STAKE

Consider using a short stake to support fruiting plants like peppers and eggplant. You can use wooden stakes, metal fence posts, rebar, or even just branches from a tree. Size the stake according to the mature height of the crop and use twine to loosely tie the plant to the stake as it grows. Since these supports are minimally invasive, you can also use them to prop up any plant that starts flopping over midseason. This sort of ongoing trellising work will help keep plants off the ground, protecting harvests and helping to keep other maintenance tasks like weeding as simple as possible. A few minutes of staking will save a lot of time down the road.

LADDER OR A-FRAME

Creating an A-frame trellis will provide a sturdy structure for heavy, vining crops. Use this type of trellis for winter squash, cucumbers, peas, or beans. A simple ladder trellis can be built by creating a rectangular frame using 2×2 lumber and attaching wire mesh or garden netting to its face. We find these trellises particularly effective for crops with aggressive tendrils. Because of the sloped nature of the structure, once the plants have located the trellis they usually form tidy and attractive plantings. Harvesting from these structures is a snap because fruits are easily visible above and below the mesh.

stem into place. Alternatively, you can get plastic or compostable vine clips, which can save tons of time if you grow a lot of trellising crops.

CAGES

Metal cages have long been used to support tomatoes, but you can also use them to trellis other crops like cucumbers, winter squash, and tomatillos. Cages come in different sizes, so be sure to check the mature height of your crop and select cages accordingly. Many gardeners prefer to build their own cages from wire-mesh fencing or other materials, creating custom structures that are often sturdier than store-bought models. The advantage of using cages is that there is virtually no ongoing training or manipulation of

TRIPOD

Tripods have traditionally been used to grow pole beans but can be adapted to support any vining crop. A tripod can be built using bamboo poles, wooden poles, plastic poles, or whatever other tall, straight material you can locate. Tripod trellises are usually made with three or four poles, but you can use as many as you like. Push the base of each pole into the soil, and use twine or wire to tie the tops of the poles together. Consider planting a fast-growing crop such as lettuce mix or arugula in the center of the tripod to utilize this otherwise wasted space. Tripod trellises are inexpensive, easy to store in the off-season, and very sturdy.

FENCE OR NETTING

Any existing or newly built fence can serve as a structure for garden plants. You can use garden netting, twine, lattice, or other material to create a surface for plants to work up. If there isn't a fence in your garden, you can use a garden net (usually made from nylon rope or plastic) to create a fencelike trellis in the garden. Netting is inexpensive and easy to set up: Simply place a support every 3 to 4 feet and tie the net to the support in as many spots as you can to ensure that it stays up. Grow peas or beans between two sets of netting or use twine to help keep plants securely on the structure as they mature. This is a particularly useful technique when you are managing large plantings of vining crops. Nets are space efficient for storage, easy to set up, and easy to manage.

HILLING

Some plants benefit simply from having soil hilled around their bases. Hilling can be especially effective for broccoli, kale, and other brassicas that have relatively shallow root systems and irregularly shaped stems. Hilling is also an effective way to keep bush beans from falling over and to bury the sprawling vines of your potatoes. Simply mound soil around the base of the plant to support the stem.

Peas and beans can be trellised with a bamboo tripod.

BEST SUPPORTS FOR INDIVIDUAL CROPS

ANNUAL VEGETABLES	SUPPORT
Basil	Hilling, single stake
Beans, bush	Hilling, single stake
Beans, fava	Hilling, single stake
Beans, pole	Cage, fence, frame, net, twine, tripod
Broccoli	Hilling, single stake
Brussels sprouts	Hilling, single stake
Collards	Hilling, single stake
Cucumbers	Cage, fence, frame, net, twine, tripod
Eggplant	Hilling, single stake
Kale	Hilling, single stake
Melon, cantaloupe, honeydew	Cage, fence, frame, net, twine, tripod
Okra	Hilling, single stake
Peas, shelling	Cage, fence, frame, net, twine, tripod
Peas, snap	Cage, fence, frame, net, twine, tripod
Peppers, hot	Hilling, single stake
Peppers, sweet	Hilling, single stake
Squash, gourds	Cage, fence, frame, net, twine, tripod
Squash, pumpkins	Cage, fence, frame, net, twine, tripod
Squash, winter	Cage, fence, frame, net, twine, tripod
Sunflower	Cage, single stake
Tomatillos	Cage, single stake
Tomatoes	Cage, fence, frame, net, twine, tripod
Watermelon	Cage, fence, frame, net, twine, tripod

PERENNIAL VEGETABLES	SUPPORT
Artichoke	Cage, single stake
Asparagus	Cage, single stake
Cardoon	Cage, single stake
Jerusalem artichoke, sunchoke	Cage, single stake

CHAPTER NINE

FERTILIZING, PRUNING & HAND POLLINATING TO BOOST PRODUCTION

Let's take a look at some of the ongoing garden management techniques that can increase your harvests and make efficient use of your time in the garden. It'll take a bit of time and practice before these skills are comfortable for you, but once you have them down, you can refine them to meet your needs. The techniques discussed here are not crop specific—you'll need to adapt them to your own crop selection and determine the right time and place to use them.

strategies for fertilizing

↓ Vegetables are needy plants. Even the lightest feeders need many more nutrients than the average ornamental plant in your landscape. Humus-rich soil and ideal conditions can supply all the nutrients vegetable crops need to grow. However, in most situations, you'll want to add supplemental fertilizers throughout the season to maximize plant health and productivity.

CROP-SPECIFIC FERTILIZING

Since conditions usually aren't ideal, you'll need to use organic fertilizer to provide the nutrients that the soil may not be able to. For example, phosphorus is harder for plants to take up when temperatures are cold, so using a high-phosphorus organic fertilizer during early-spring planting will help ensure the plants have all they need.

Side-dressing onions with blood meal gives them a boost of nitrogen.

If you're growing multiple crops in one plot over the course of the season, or growing heavy-feeding crops, it can be very difficult to keep nutrient levels high enough for good production without the addition of organic fertilizer. Heavy feeders such as tomatoes, cucumbers, and winter squash often need more nitrogen than is available in the soil, so a high-nitrogen organic fertilizer will help keep these crops vigorous. As you become more experienced and fine-tune your soil management, you may be able to minimize or eliminate your use of additional organic fertilizer. In most cases, it will remain an important part of your soil fertility plan.

USING SLOW-RELEASE & SOLUBLE FERTILIZERS

Generally speaking, the nutrients in organic fertilizers are not immediately available to your crops—they must be broken down by soil bacteria first and become available slowly over a period of time. Because of this, organic fertilizers are often referred to as insoluble or "slow release." The slow breakdown is one of the great benefits of organic fertilizers, especially granular products. There is less risk of losing soluble nutrients to leaching or off-gassing, and the nutrients are supplied to the plant at a rate that they can actually absorb and use.

GIVING PLANTS A QUICK BOOST

Slow-release fertilizers are great, but plants often need a quick boost if they're experiencing nutrient deficiencies. Water-soluble nutrients are available almost immediately. Slow-release fertilizer is like feeding your plants a steak;

water-soluble fertilizer is like an energy bar in the middle of a demanding workout. Using liquid organic fertilizer is a good way to supply soluble nutrients to plants in need. Water-soluble nutrients can help reduce transplant shock at planting time and can help quickly turn around a crop that is showing visible signs of nutrient deficiency. If a long-season crop like a tomato or squash is stunted or showing discoloration, the addition of a liquid fertilizer can save the planting, getting the crop back on track for an abundant harvest. Short-season crops may be less salvageable once showing visible signs of nutrient stress, but scheduled applications of liquid nutrients can help prevent these issues in virtually any annual crop.

Blood meal can also supply a shot of soluble nitrogen, along with a longer-lasting supply of insoluble nitrogen. Blood meal can contain anywhere from 1 to 6 percent soluble nitrogen, depending on how it's processed. It's important to remember that you can't supply all of the nutrients your crops need in a soluble form at one time, so make sure your main priorities are to maintain soil quality with compost and use granular fertilizer before planting.

DIY ORGANIC FERTILIZERS

You can make your own organic fertilizers by mixing different ingredients together to create your desired N:P:K ratio. We've included a few of our favorite recipes here; you can find many others online or in other books. When mixing the fertilizer, be sure to do it outside and wear a dust mask. Use a 5-gallon bucket and a trowel for mixing smaller quantities, and a clean wheelbarrow and a shovel for larger quantities.

GENERAL PURPOSE FOLIAR FEED

If fish emulsion is used, this foliar feed primarily provides nitrogen. The N:P:K varies depending on the breakdown of the fish emulsion you choose.

× 1 gallon water

× 3 tablespoons fish emulsion (omit for plants that are setting fruit)

× 3 tablespoons liquid kelp

× 2 tablespoons molasses

GENERAL PURPOSE MIX
(N:P:K approximately 4:4:3.5)

× 3.25 parts blood meal

× 1 part bone meal

× 2.25 parts greensand

× 1 part kelp meal

HIGH-PHOSPHORUS TRANSPLANTING MIX
(N:P:K approximately 4:6:2)

× 3.25 parts blood meal

× 1.5 parts bonemeal

× 2.25 parts greensand

× 1 part kelp meal

VEGAN FERTILIZER
(N:P:K approximately 3:3.1:3.1)

× 0.25 part greensand

× 0.5 part kelp meal

× 0.1 part potassium sulfate

× 1.5 parts alfalfa meal

× 4 parts cottonseed meal

× 1 part bat guano

GERMINATION FERTILIZER
(N:P:K approximately 3:3:2)

× 1 part blood meal

× 0.75 part bonemeal

× 1.25 parts greensand

× 0.25 part kelp meal

READING THE LABEL

The major nutrients plants require for good growth are nitrogen, phosphorus, and potassium. These are often called macronutrients and abbreviated as N:P:K. Calcium and magnesium are also highly important, as are a host of important micronutrients and trace minerals, such as sulfur, boron, zinc, manganese, iron, and copper.

Every fertilizer is labeled with its N:P:K ratio. For example, a fertilizer that contains 5 percent nitrogen, 5 percent phosphorus, and 5 percent potassium by weight will be labeled 5:5:5. Typically, several ingredients are mixed together to create the desired N:P:K ratio.

Sometimes calcium, magnesium, and other micronutrient amounts are also labeled on the package, but the three numbers together always refer to nitrogen, phosphorus, and potassium. If you want to keep things simple, you can use a "balanced" fertilizer for all your pre-planting and side-dressing needs. Balanced means that all three numbers are nearly the same—for example, 5:5:5 or 3:2:3—which in turn means the fertilizer supplies approximately equal amounts of nitrogen, phosphorus, and potassium to your plants.

- × **Nitrogen** is responsible for early vegetative growth in plants.
- × **Phosphorus** is crucial for photosynthesis, early root growth, and flowering.
- × **Potassium** helps with photosynthesis, fruit formation, and disease resistance.
- × **Calcium** is vital for cell structural development, root growth, and nitrogen uptake.
- × **Magnesium** plays an important role in photosynthesis, and must be present in a 1:5 to 1:7 ratio with calcium for both nutrients to be available to crops.

FOLIAR FEEDING

Surprisingly, plants can take up nutrients very quickly and effectively through the stomata (pores) on their leaves. In fact, stomata can absorb nutrients more effectively than roots. However, it's not possible for plants to take up large amounts of nutrients this way, so foliar feeding is best used when you want to give your crops a quick boost. Many growers apply foliar sprays at the onset of fruiting. Fruiting requires significant nutrient deployment, and a quickly absorbable feed can help keep the plant healthy and improve overall yields. The timing of foliar feeding applications is crucial: Applying foliar feeds in full sun can be damaging to the plant. We suggest waiting for a cloudy day, or applying sprays before eight o'clock in the morning.

Almost any organic liquid fertilizer can be used for foliar feeding, but it must be diluted as labeled to prevent burning the foliage. (See our General Purpose Foliar Feed, page 139.) The fertilizer should be applied to the leaves in a fine mist. A hand pump sprayer such as those made by Solo work well for small applications; backpack sprayers are more efficient for spraying a larger space.

Apply the diluted fertilizer until you just start to see liquid dripping from the leaves. Adding a surfactant such as molasses or yucca extract will help the fertilizer adhere to the plant and give it time to absorb all of the nutrients. Most vegetable crops have the majority of their stomata on the underside of the leaf, so spraying foliar feeds in this area will maximize absorption of the nutrients. This rule covers all dicotyledon plants (those with two cotyledons or seed leaves). Monocots (plants with a single cotyledon) have equal numbers of stomata on the top and bottom of the leaf. Corn, rye, and other grasses are monocots.

WHEN TO APPLY FERTILIZER

We recommend applying organic fertilizer as follows.

Before planting. Add granular/dry organic fertilizer to the soil prior to planting any vegetable crop. For direct-seeded and closely spaced transplanted crops, broadcast it over the surface of the soil and mix it into the top 4 inches before planting. For widely spaced crops like tomatoes, potatoes, or winter squash, mix it into each hole before planting.

After planting. Side-dress heavy-feeding crops with granular fertilizer or liquid fertilizer, or apply a foliar feed, at 2, 4, and 6 weeks after planting. If using granular fertilizer, sprinkle the recommended amount around the base of the plant, mix it into the top inch of the soil, and water it into the soil. If using liquid fertilizer, measure the desired amount into a watering can or hose-end sprayer and dilute it as recommended on the label. Make sure to rinse out the can or sprayer after each use, as organic liquid fertilizer will stink and clog up screens and outlet holes.

Here are a few suggestions for how to fine-tune your fertilizer applications.

+ Apply low doses of liquid fertilizer to your nursery transplants two to three times per week. Applications can begin as soon as seedlings have emerged and continue until they're transplanted into the garden.

+ When transplanting, use a high-phosphorus fertilizer (2:4:2 or something similar) to encourage quick root development.

+ Apply a foliar feed with lots of phosphorus, potassium, and calcium (0:5:5 or similar) at the onset of fruiting to improve fruit quality.

+ Side-dress leafy and brassica crops with a high-nitrogen fertilizer such as 5:1:1. This works especially well for leaf crops, which

In most situations, you'll want to add supplemental fertilizers throughout the season to maximize plant health and productivity.

don't require large amounts of phosphorus and potassium. We use straight blood meal (12:0:0) for side-dressing garlic, onions, brassicas, arugula, and lettuce.

WHAT'S IN ORGANIC FERTILIZER?

Organic fertilizers are made from a variety of ingredients derived from mineral, animal, and plant sources. Mineral fertilizer sources include greensand, rock phosphate, sulfate of potash, and glacial rock dust. Generally, animal by-product ingredients come from livestock operations or fisheries. Common animal waste products include blood meal, cow or fish bone meal, feather meal, crab meal, whole fish meal, bat guano, chicken manure, and worm castings. Animal by-product–based ingredients are processed to kill pathogens and remove impurities. Common plant waste fertilizers include cottonseed meal, soybean meal, kelp meal, and alfalfa meal.

WHICH CROPS NEED A BOOST

We recommend supplemental fertilizing for heavy feeders or plants that need extra nutrition early on in life. Frequency of supplemental fertilizing will vary depending on whether you're applying granular, liquid, or foliar fertilizers. Soil and climate conditions also affect application frequency. Foliar feeding or liquid fertilizing is often carried out every week or every other week. If you're side-dressing with granular fertilizer, one to three applications are usually sufficient. Keep in mind that any crop may benefit from supplemental fertilization, especially those that show signs of nutritional deficiency. Whether or not they need additional fertilizer during the growing season, all the annual crops listed below should be fertilized at planting time. We also recommend fertilizing perennial vegetables and fruits before planting, and again in the spring of each growing season. Perennial herbs should be fertilized at planting time, but may or may not need a boost in successive seasons.

ANNUAL VEGETABLES & HERBS THAT NEED SUPPLEMENTAL FERTILIZING

- Beets
- Broccoli
- Brussels sprouts
- Cabbage
- Cabbage, Chinese
- Cauliflower
- Corn, sweet
- Cucumbers
- Eggplant
- Garlic (in spring)
- Melon, cantaloupe, honeydew
- Okra
- Onions, bulb (before bulbing initiates)
- Peppers, hot
- Peppers, sweet
- Potatoes
- Squash, gourds
- Squash, pumpkins
- Squash, summer
- Squash, winter
- Sweet potatoes
- Tomatillos
- Tomatoes
- Watermelon

ANNUAL VEGETABLES & HERBS THAT DON'T NEED SUPPLEMENTAL FERTILIZING

- Arugula
- Basil
- Beans, edible soy (edamame)
- Beans, fava (broad)
- Beans, lima
- Beans, shell
- Beans, snap
- Broccoli raab
- Carrots
- Celeriac
- Celery
- Chard, Swiss
- Cilantro
- Collards
- Dill
- Endive
- Fennel, bulbing
- Kale
- Kohlrabi
- Leeks
- Lettuce, baby mix
- Lettuce, head
- Mâche
- Mustard greens
- Pak choi
- Parsley
- Parsnips
- Peanut
- Peas, shelling
- Peas, snap
- Radicchio
- Radishes
- Rutabagas
- Scallions
- Spinach
- Turnips

Cutting off the top of a Brussels sprout plant 3 to 4 weeks before harvest will increase the size of the sprouts.

Late-season tomato shoots won't have time to produce mature fruit; they can be removed to prevent the plant from wasting energy.

pruning for production

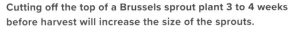 Pruning—the practice of removing roots, stems, branches, leaves, flowers, or fruits from a plant—can serve many goals. It can increase a plant's productivity, speed up ripening, reduce the chance of pest and disease issues, prevent a plant from overwhelming and shading out neighboring crops, and simply keep the garden organized and tidy.

PRUNE FOR AIR CIRCULATION & PLANT HEALTH

Some crops will simply grow too vigorously for their own good (or at least for your own good). Plants can put on a tremendous amount of foliage during the course of a season. Under certain circumstances, this extra vegetative growth may reduce overall yield. As an example, indeterminate tomato plants will continually produce new growth shoots from their leaf axes. Removing new supplemental shoots during the fruiting period will help prevent the plant from wasting energy developing new branches that will not have time to ripen fruit before the season's end. By pruning, you can help the plant prioritize where resources are allocated.

Dense foliage can also create conditions for an influx of a pest or disease. Good air circulation around plants helps dry out wet leaves and stems, which reduces sites for fungal and bacterial infection.

Overgrown plants are harder to manage and you're more likely to accidentally break off branches or otherwise damage an unwieldy plant.

Removing immature flowers will help keep a leafy plant like kale in production longer. The flower shoots also provide an excellent secondary harvest.

Accidental breaks will create sites for disease entry, so it's better to intentionally prune the plants to a manageable size and shape to keep it healthy. It can also help simplify fertilization and harvesting.

STRATEGIC BLOSSOM REMOVAL

Believe it or not, you can dramatically affect the productivity of crops by managing the timing of their blossoms. For every plant, there is a sweet spot during its life cycle when the development of flowers is most likely to result in the highest yield. If a fruiting crop flowers too early, its growth can be stunted, resulting in little to no viable harvest. If a crop continues to flower as the season is drawing to a close, it wastes energy that would be better diverted to existing underripe and developing fruit.

HARVEST BOLTING PLANTS

Bolting is simply a term for undesirable flowering—especially in root, stem, or leaf crops. It's usually a response to stress from unfavorable weather, lack of water, lack of nutrients, or other environmental stresses. Flowering takes a lot of energy, so once the plant begins to invest energy in flowers, the growth of the roots, stems, and leaves of the plant will decrease or stop entirely. It's common to find that a bolting vegetable has begun to taste bitter, as it's stopped supplying nutrients and sugars to the nonflowering parts of the plant.

As a general rule, once the flowering mechanism inside the plant has been tripped, it is very difficult to prevent it from flowering again. Cutting back a bolting flower stalk will usually result in a new flower stalk emerging within a few days. In most cases, once a plant starts to bolt, the best thing you can do is to harvest the crop and prepare the space for a new planting. (Unless, of course, you've chosen to leave the bolting plant in place in order to collect seed or attract beneficial insects.)

HOW TO PREVENT OR DELAY BOLTING

To delay bolting, reduce the plant's stress level as much as possible by growing it during the season it prefers, keeping it evenly and adequately watered, and providing it with nutrient-rich soil. It's even possible to select varieties of a crop that have been bred specifically to bolt more slowly. Even in the best circumstances, some crops will bolt. Among a healthy row of carrots or leeks, you may come across a few rogue, bolting individuals. Remove erratically bolting crops from healthy plantings as soon as you notice them. Leaving them in place will take up space, water, and nutrients that would be better utilized by healthy plants.

Root pruning can encourage a tomato to ripen its fruit earlier.

REMOVING EARLY BLOSSOMS

Ideally, plants shouldn't set blossoms until they're large and established enough to ripen good yields of quality produce. If fruiting plants are allowed to bloom and set fruit when they're very small, it will inhibit them from growing a large root system and strong branches for the fruits. Early fruiting will also limit the overall size, yield, and quality of fruits.

If a transplant has become stressed during its nursery grow out, it may begin to set blossoms even while in a small pot or shortly after transplanting. This often happens when a plant is kept in its container for too long. If you notice early blossoms, simply pinch them off. Pinching back early blossoms for a few weeks while the plant becomes established will enable it to grow to its full potential.

REMOVING LATE BLOSSOMS

As the season winds down, fruiting crops may continue to flower late into the season. Indeterminate tomatoes and pole beans won't necessarily recognize that the season is coming to an end, and will keep producing flowers in the expectation they will have time to mature into fruits with viable seeds.

Knowing your average first frost date, you can anticipate how many weeks of useful production these crops might have left and begin removing any new flower blossoms several weeks before the expected change in weather. By removing blossoms that have no hope of producing usable fruit, you can encourage the plant to ripen its existing fruits more quickly, resulting in a better quality late-season harvest.

DISCOURAGING FLOWERS ALTOGETHER

Fruiting and flowering crops such as tomatoes, peppers, squash, and sunflowers must bloom in order to produce a harvest. However, for crops that are harvested for their roots, leaves, or stems—such as carrots, celery, or lettuce—flowering generally signals the end of the plant's usable harvest. You'll want to actively discourage these crops from flowering by regularly cutting or pinching off any emerging flowers.

ROOT PRUNING

Root pruning is a simple technique that can help speed up the ripening process of crops; it is often employed to help spur quicker ripening in tomatoes and other long-season fruiting crops. Use a shovel or hand tool to sever the root system of the plant in a semicircle around its base. Cutting the roots shocks the plant into emergency mode, speeding up the ripening process as the plant attempts to create viable seeds before its imminent demise.

Pinching basil regularly will produce bushier, more productive plants.

Cutting the tops of peas can encourage the plant to flower and produce more mature fruit lower on the vine.

PINCHING FOR HIGHER YIELD

Some plants will produce much higher yields if pinched back to encourage branching. Particularly effective on basil and related herbs, pinching can dramatically increase your harvest. Use your fingers, scissors, or pruners to snip off the top set of leaves of each branch of the plant—be sure to snip all the way down to the next set of leaves on the stem. The plant will then send out two new branches from those leaf axes. When these new branches have grown to harvestable size, pinch them both back and double your harvest again.

PRUNING TO CONTAIN VINING CROPS

Trellised vining crops, such as peas, beans, or vining squash, can be maintained at a particular height by cutting back the primary growth tip (also called the apical meristem). This can be effective if you want to prevent the plant from growing past the top of a trellis. Simply use your scissors or hand pruners to snip off the top of the plant at the desired height. It may be necessary to cut back this top growth several times through the season to maintain this height.

hand pollinating for higher yields

You may find it desirable—or even downright necessary—to hand pollinate your plants. Pollinating by hand can boost yields of fruiting crops. It's especially helpful if you have a dearth of pollinating insects in your area, if you're growing fruiting crops under cover where insects may have a difficult time reaching them, or if you're selecting plants for seed saving. Cucurbit crops, in particular, benefit from hand pollinating, as they're the most likely to suffer from low insect pollinator populations.

The process is simple. Most gardeners will use a small paintbrush or cotton swab as their pollinating tool, but you can also just use your finger (with or without a glove). Another option is to simply cut or pinch off a few flowers to use as your handheld pollination tool. The goal is to move pollen from the male part of the flower to the female part of another flower.

In situations where there is a large crop and/or a crop with many flowers, it's possible to help encourage pollination simply by shaking the plant to release pollen. Commonly done with greenhouse tomatoes and sweet corn, this technique can be useful on a range of fruiting crops. When flowers are open and fully developed, gently shake the plants to release pollen into the air and to scatter it across the plants. Some of the airborne pollen will land on receptive flower stigmas, increasing the number and quality of fruits.

CUCURBIT CROPS BENEFIT FROM HAND POLLINATING, AS THEY'RE THE MOST LIKELY TO SUFFER WHEN INSECT POLLINATOR POPULATIONS ARE LOW.

HOW TO HAND POLLINATE

To achieve successful pollination, it is crucial that you can identify the parts of the flower and determine what kind of flowers you have. Some crops produce separate male and female flowers (these plants are called monoecious), others have both sexes on a single flower (hermaphroditic, or "perfect"), and some even have separate male and female plants (dioecious).

1 Determine whether the crop is hermaphroditic, monoecious, or dioecious.

2 Locate the stamens and the stigma of the flowers.

3 Take your chosen tool (a paintbrush can be used for many crops) and rub it against the stamens of the flower.

4 Transfer this pollen-laden tool to the stigma of the same flower or to another flower (depending on the type of flower) on the same plant or a nearby plant of the same variety (depending on the crop).

5 Repeat and wait for fruits to develop.

GLOSSARY OF FLOWER TERMS

- × **Complete.** Flowers containing sepals, petals, stamens, and pistil
- × **Incomplete.** Flowers lacking sepals, petals, stamens, and/or pistil
- × **Perfect.** Flowers containing male and female parts
- × **Imperfect.** Flowers lacking either male or female parts

- × **Pistillate.** Flowers containing only female parts
- × **Staminate.** Flowers containing only male parts
- × **Hermaphroditic.** Plants with perfect flowers
- × **Monoecious.** Plants with separate male flowers and female flowers on the same plant

- × **Dioecious.** Plants with male flowers and female flowers on separate plants
- × **Gynoecious.** Plants with only female flowers
- × **Andromonoecious.** Plants with only male flowers

ANATOMY OF A FLOWER

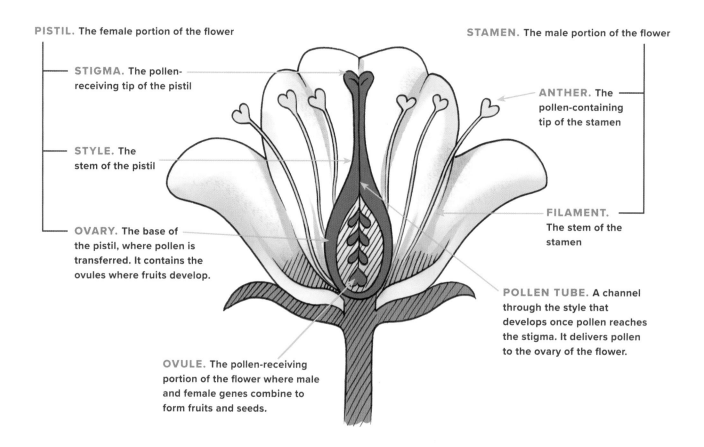

PISTIL. The female portion of the flower

STIGMA. The pollen-receiving tip of the pistil

STYLE. The stem of the pistil

OVARY. The base of the pistil, where pollen is transferred. It contains the ovules where fruits develop.

OVULE. The pollen-receiving portion of the flower where male and female genes combine to form fruits and seeds.

STAMEN. The male portion of the flower

ANTHER. The pollen-containing tip of the stamen

FILAMENT. The stem of the stamen

POLLEN TUBE. A channel through the style that develops once pollen reaches the stigma. It delivers pollen to the ovary of the flower.

FLOWER TYPES FOR POLLINATING

When hand pollinating, use this chart to determine if the crop in question is hermaphroditic or monoecious.

ANNUAL VEGETABLES & HERBS	FLOWER TYPE	SELF-POLLINATING?
Beans, edible soy (edamame)	H	Yes
Beans, fava (broad)	H	Yes
Beans, lima	H	Yes
Beans, shell (fresh or dried)	H	Yes
Beans, snap	H	Yes
Corn, sweet	M	Yes
Cucumbers	M	Variety dependent
Eggplant	H	Yes
Melon, cantaloupe, honeydew	M	Yes
Peas, shelling	H	Yes
Peas, snap	H	Yes
Peppers, hot	H	Yes
Peppers, sweet	H	Yes
Squash, gourds	M	Yes
Squash, pumpkins	M	Yes
Squash, summer	M	Yes
Squash, winter	M	Yes
Tomatillos	H	No
Tomatoes	H	Yes
Watermelon	M	Yes

H = Hermaphroditic M = Monoecious

CHAPTER TEN

MANAGING PESTS & DISEASES

Regardless of the precision and care you put into crop planning and management, Mother Nature will inevitably come knocking at the garden gate looking for a piece of the action. Just as you can improve your yields by learning more about the life span and growth habits of your crops, you can improve yields by learning about the lives of the pests and diseases that prey on them. Unmonitored or unmanaged pest or disease issues can dramatically decrease your produce quality and overall garden productivity. It's important to apply the same amount of forethought and attention to pests and diseases as you've invested in every other part of the garden.

basic concepts of pest & disease prevention

Bear in mind that it's simply not possible to eliminate all pests and diseases from the garden. The goal is to manage the problem, not to solve it. Organic gardening is a way of communicating and engaging with the greater environment, not controlling it. Thus, pest and disease management techniques should disrupt the natural environment as little as possible, and should cooperate with and simulate natural processes as much as possible. Growing food is a humbling and profound meditation. The next time you're struggling with a difficult pest issue, try to appreciate the awesome power of nature and your small place within it.

We believe in starting with preventive and minimally invasive practices and ramping up the methods if specific problems get out of hand. Maintaining a healthy environment and keeping a watchful eye can help you avoid many problems before they start.

START WITH HEALTHY SOIL

It's simple: Healthy, biodiverse soil produces healthy plants, and healthy plants have more resources to fight off pests and diseases. So, if you're following the right steps to build and maintain your soil (see Chapter 5), you're off to a great start.

BE VIGILANT

Pest and disease management, like all aspects of garden management, is much easier and more effective when it happens frequently. Make sure to visit your garden every day, even if you only have a few minutes to walk through the beds and look at the crops. If you're constantly inspecting the garden, you'll notice problems early. Early detection gives you the opportunity to research the issue and find a solution before things get out of hand.

Vegetable pests and diseases spread very quickly. Just as your plants must grow and reproduce in a very short period of time, so do the pests and diseases that feed on them. Many insect pests will produce several generations in a single season, sometimes producing an entire new generation in a matter of a few days or weeks. These populations expand exponentially unless you or some other natural force is there to keep them in check. If a pest population has weeks of time to establish itself before you recognize it, it will have already spread beyond its initial holdfast and will be much more difficult to eradicate.

Slugs are fond of wet, tender leaves like those found on heads of lettuce.

UNDERSTAND YOUR CROPS

The longer you tend a garden, the faster and easier it will become to identify problems. You'll also come to understand that each crop you grow, and even particular varieties of a crop, will have their own tendencies and peculiarities.

For example, on a hot day you might wonder if the leaves on your broccoli plants are wilting because the plants need water or because they're being attacked by a pest. Upon inspecting the soil, you find that it's adequately moist. After some research and a discussion with a local expert, you learn that brassica leaves tend to droop on very hot days, usually when the temperature is above 90°F (32°C). You keep a close watch on the plants, and everything seems to return to normal when the temperature drops a bit. You're relieved, because the problem could have been root maggots eating your crop from underground. It is this kind of assessment and follow-through that will build your knowledge and help keep your garden thriving.

Even when they have adequate moisture, broccoli leaves will sometimes wilt; it's a protective response during periods of high temperatures.

FINDING HELP WITH PEST & DISEASE IDENTIFICATION

Use whatever resources you have to proactively learn what problems are most prevalent in your area. A quick internet search for "most common vegetable pests and diseases in [your area]" will yield a great amount of information. The detail and accuracy of information on the web is not always consistent or reliable, so it is usually a good idea to check numerous sites and to talk with local gardeners and experts.

University Extension agencies are often still the best and most knowledgeable on local issues and can even help identify pests and diseases for you. Depending on the school and the particular person, their suggested remedies may not always be appropriate for your garden. Some Extension offices are proponents of chemical pesticides and fungicides, but at the very least they can help you figure out what your problem is. Fortunately, many universities are beginning to research and develop support for organic agriculture. Your interaction with the agency might just help push them in this direction, so don't be afraid to reach out.

Local garden nurseries and garden clubs can be great sources of support, as can investing in a few robust pest and disease books for your gardening library. Somebody out there has had the same issue that you are experiencing. It might be very easy to find a satisfactory solution, or it may take some sleuthing. At the very least, you'll learn a lot during each research session.

take preventive action

↓ Knowing the most prevalent pests and diseases in your region will allow you to prepare for their attack and minimize or eliminate the problem before it happens. The more you understand about pest and disease life cycles, the better prepared you'll be to combat them. For example, flea beetles and potato beetles can emerge early in spring. You may be able to reduce their population by waiting a few extra weeks to plant out their favorite crops, effectively starving these pests-in-waiting and experiencing much lower pest pressure for the rest of the season. Strategies vary, depending on the issue at hand, but there are a few highly effective and widely adapted preventive techniques that we think you'll find useful.

ATTRACT BENEFICIAL INSECTS

The presence of insects in the garden can be a good or a bad thing. It depends on the insect. While some insects might nibble on your crop plants, others will prey on pests and help pollinate your plants. In the garden, helpful insects are called beneficials. Because the presence of insects is inevitable, the best strategy is to help build a diverse ecosystem in and around your property. Just like in a natural ecosystem, if you can achieve high levels of biodiversity, the population of any particular insect will be kept in check by the presence of other insects that prey on and parasitize them.

Sunflowers are beautiful, produce a delicious harvest, and attract beneficial insects.

This polyester row cover allows sunlight to reach your crops and will exclude pests such as carrot rust fly.

Add flowers. Planting a range of flowers in the yard is the easiest and best way to create habitat for a buzzing metropolis of insect life. You can add small, fast-growing, flowers like alyssum to the edge of any garden bed and mix taller annuals like cosmos, bachelor's button, and love-in-a-mist in between edible crops. You can create a dedicated space in the garden for perennial flowers like echinacea, cardoons, bee balm, and helenium.

Let a few crops bolt. Another strategy is to allow certain crops or a portion of a planting to flower instead of removing them promptly after primary production has finished. Crops like cilantro, dill, fennel, basil, arugula, and mustard greens will produce flowers that beneficial insects love. Many of these flowers can also be eaten on salads or in recipes. Cilantro, if left to generate seeds, will supply you with coriander.

PHYSICAL EXCLUSION

Putting up a physical barrier between the pest and the crop is often the most effective strategy for reducing pest pressure on your plants.

Floating row covers. When properly used, floating row covers make it nearly impossible for many pests to munch on or lay eggs near your

FLOWERS THAT ATTRACT BENEFICIAL INSECTS

- Alyssum
- Arugula
- Bachelor's button
- Basil
- Bee balm

- Buckwheat
- Cardoon
- Chamomile
- Cilantro
- Cosmos
- Dianthus

- Dill
- Echinacea
- Fennel
- Helenium
- Hyssop
- Lavender

- Love-in-a-mist
- Marigolds
- Mint
- Mustard greens
- Parsley
- Sunflower

- Valerian
- Verbena
- Zinnia

crop. Since these covers prevent insects from entering the bed, it's important to know the life cycle of the pest to determine when to apply them. This is often immediately after seeding or transplanting. Depending on the pest and the crop, covers may be left on for a few weeks while the plant establishes itself, or all the way up to harvesting time. If you're growing a fruiting plant that needs insect pollination, you'll need to remove the cover at flowering.

If the pests you're fighting don't overwinter in the soil, then a cleared, prepared bed can be planted with a crop and covered to keep pests out. However, if you have a pest that overwinters in the soil between seasons, a row cover may trap the emerging insects right in the bed with your crops.

Row cover is especially helpful to control caterpillars, beetles, stink bugs, root maggots, and borers. Plants that benefit most from row cover will vary depending on what pests are most problematic in your area. For the most effective exclusion of pests, cover the crop immediately after seeding or transplanting.

Row covers used for season extension, typically made from spun-bonded polyester, also work well for pest exclusion. Row cover fabric comes in a variety of different weights. Light fabrics let in more light and trap less heat so summer crops are less likely to overheat when temperatures are high. Heavier fabrics provide better protection from frost but can reduce photosynthesis in your crops. Some very lightweight fabrics made from fine mesh are designed specifically for pest control rather than as a broad-spectrum season-extension tool. These allow more air circulation and light penetration than polyester row covers.

Netting. Flexible polyethylene netting, called bird netting, can be used to protect crops from hungry birds and other critters. Bird netting can be attached to hoops, like floating row cover, or it can simply be draped over the crops. Birds are especially attracted to sweet-tasting crops like

These kale plants are largely free of insect damage thanks to the use of floating row cover.

Once you've identified the problem you're dealing with, you may find creative and surprising opportunities for pest-specific physical exclusion. Simply taking the time to learn about the pest's life cycle and habits might help you find a way to easily exclude it from your crops.

For example, root maggots can be a scourge on brassica crops. The maggots are hatched from eggs laid at the base of a broccoli or kale plant, where they burrow underground and feed on the roots of the crop. The adult form of the pest, a small fly that closely resembles the common house fly, prefers to lay eggs right next to the target plant. Knowing this makes it possible to cover this area with an impervious material like plastic mulch, cardboard, or cut-up pieces of landscape fabric. These materials prevent the fly from laying its eggs at the base of the plant.

corn and strawberries, but are also known to eat seeds right out of the soil and tear out young transplants, which then dry out and die in the sun.

You don't need to leave bird netting on crops continuously; you can simply install it when the crop is at its most vulnerable. Drape the netting over newly seeded crops until they've germinated and emerged from the soil. For sweet-tasting crops, use the netting right before fruits begin to ripen and remove it when harvesting is finished for the season. If transplants are being attacked, apply the netting at planting time and leave it in place until the plants have become established, usually a week or two.

Fencing. Fences are an excellent solution for keeping larger mammals out of the garden. The most common garden fencing is built to exclude deer and other large herbivores. A deer fence should be at least 8 feet tall to keep deer from jumping into the garden. Deer fences provide an easy opportunity to exclude any other, smaller animals as well. By including a fine wire mesh, like chicken wire, on the bottom 3 feet of the fence, it's possible to keep out rabbits and other small mammals. If tunneling pests are a problem, the fencing should be buried 1 to 2 feet underground around the perimeter of the garden.

PREVENTIVE APPLICATIONS

Some organic sprays and biological controls are most effective if applied before problems have begun. In order to use your time and money wisely, you'll only want to use these methods for pests that are known problems in your area or that you've had issues with in the past.

Nematodes. These microscopic creatures live in the soil and feed on insects, including fungus gnats, thrips, leaf miners, onion maggots, root maggots, caterpillars, cutworms, and armyworms. There are many different species of nematodes, and not all of them are beneficial to the garden. Each type will target specific insects, so a little research should reveal the appropriate nematode to apply for the pest you're dealing with. It's also possible to purchase packages of beneficial nematodes that include several different species, so you can apply a range of these helpful parasites and have a better chance of keeping pest species at bay.

Nematodes will generally come in a plastic container that should be kept refrigerated until use. The dormant creatures will be hibernating in a medium that you add to water and soak into the soil. Because nematodes need a wet environment to survive, it is best to apply them when the soil is already moist. Plan to irrigate or wait for rainy weather to soak the soil before applying them.

Applied nematodes are generally very short lived, which is a good thing ecologically speaking, as it prevents these critters from becoming invasive. However, this also means that you may need to apply them several times a season for maximum effect.

Diatomaceous earth. This is a very strange but effective organic pest deterrent. Diatomaceous earth is a white or off-white powder made from the fossils of very tiny, single-celled organisms called diatoms. Although the powder feels relatively smooth to your hands, it has microscopic jagged edges that can slice open soft-bodied insects like slugs and snails. Be careful not to breathe the dust, as it can be damaging to your lungs. Wear a dust mask and safety goggles when applying it. Diatomaceous earth can be mixed into the garden soil to deter soil-dwelling larvae like root maggots, or it can be applied to the surface of the soil to keep slugs and other surface-dwelling insects at bay.

DIY PREVENTIVE SPRAYS

There are many approved organic pesticides and fungicides that can be applied in spray form. Even though they're organic, the more potent sprays should only be used as a treatment for an existing problem to avoid the possibility of harming beneficial insects. However, there are some mild sprays that can be used preventively to great effect to minimize the establishment of pests and diseases. We often use a baking soda spray preventively for fungal issues and a hot pepper and garlic spray to prevent insect and animal pest damage.

BAKING SODA SPRAY

Use this spray as prevention or as treatment at the first sign of a fungal disease such as powdery mildew, downy mildew, or damping-off.

- × 1 tablespoon baking soda
- × 1 tablespoon vegetable oil
- × 1 gallon water
- × ½ teaspoon liquid dishwashing soap (or castile soap)

Mix the baking soda and vegetable oil with the water in a sprayer. Shake it up thoroughly and add the soap. Be sure to agitate your sprayer while you work to keep the ingredients from separating.

Spray your plants weekly, preferably on overcast days to prevent the spray from burning the foliage. Cover upper and lower leaf surfaces and spray some on the soil. It won't get rid of the fungus on leaves that already have it, but it will prevent the fungus from spreading to the rest of the plant.

HOT PEPPER & GARLIC SPRAY

The oils from hot peppers and garlic can deter birds, mice, rabbits, dogs, and a range of insect pests. Spray preventatively or at the first signs of damage.

- × 6 cloves garlic
- × 1 small onion
- × 1 teaspoon ground cayenne pepper (or 6 hot chile peppers)
- × 1 quart water
- × 1 tablespoon liquid dishwashing soap (or castile soap)
- × 1 tablespoon vegetable oil or mineral oil (not olive oil)

Chop up or use a blender or food processor to mince the garlic, onion, and cayenne. Add to the water. Steep the mixture for several hours or overnight in the refrigerator. Strain through cheesecloth, add the soap and vegetable oil, and shake until well mixed.

Using a spray bottle, spray to cover your plants, including leaf undersides and on soil under plants. You can also apply it to the ground around your garden perimeter to discourage larger animal pests. Store in the refrigerator in a labeled spray bottle or jar for up to a week.

Note: Peppers can cause skin and eye irritation. Be careful, wear gloves, and don't touch your eyes after handling the spray. When harvesting produce that has been sprayed with hot pepper and garlic spray, wash it extra carefully.

take direct action

Sometimes preventive techniques just don't work and you need to take immediate direct action to get a situation under control. In many cases, you can address the problem simply by removing pests or diseased plants from your garden. These techniques are effective in small areas or when curtailing a problem early on, but lose their appeal quickly in larger gardens.

HAND PICKING

Hand picking insect pests might seem gross, but becoming closely acquainted with your adversaries in this way will teach you a lot about their preferences. You'll see if they hang out on the tops of leaves or bottoms of leaves, new leaves

Regular scouting for insects allows you to discover problems like imported cabbageworm before they get out of hand.

or old leaves, broccoli or cauliflower. Over time you'll be able to quickly locate pest problems. In the event you need to graduate to a more serious management strategy, you can be more selective about the application of organic sprays.

It may take some time to recognize the pest you're picking. As an example, noticing green caterpillars such as cabbage loopers or imported cabbageworms on your brassicas may be very challenging at first. Once your eyes learn to scan the plants for the pest's particular color and shape, you'll find it increasingly simple.

When dealing with small pests such as aphids, simply squash them with your fingers rather than attempt to remove them from the plant. Running your fingers along a stem or leaf with slight pressure will allow you to kill hundreds of these tiny pests without sprays or any other supplies.

Hand picking can also be helpful in preventing the spread of disease. By removing the leaves or portions of the plant that are most affected, you may be able to slow down or eliminate the spread of the disease without the use of sprays.

If a pest or disease is heavily attacking a small group of entire plants, consider sacrificing these plants for the greater good of the garden. Often, pests and diseases first appear on the weakest plants in the group. Quickly removing these plants will also remove the pests and the potential subsequent generations of these pests.

TRAPS

Sticky traps can help reduce pest pressure with minimal management time. The most common garden traps are sticky yellow cards placed throughout the garden and/or nursery space. These can be purchased or you can make your own by applying a horticultural pest glue, such as Tanglefoot, to bright yellow paper, plastic,

Sticky traps can be used to catch a wide number of pests in the garden.

WHAT TO DO WITH DISEASED PLANTS

Although the composting process can be very effective at killing fungal and bacterial pathogens that cause disease, most growers agree that diseased garden plants should not be added to home compost piles or worm bins. The risk of reintroducing pathogens to the garden is too great.

If you're lucky enough to have a municipal composting system, you can put these plants in your yard waste bin or take them to the nearest collection facility and have them composted professionally. Commercial composting systems are designed to reach higher temperatures than home systems can typically achieve, so plant pathogens are more effectively killed.

Another option for home growers is to burn diseased plants, or if you have woodland or unused space on your property, you can bury diseased plants away from the garden site. If none of these options are feasible, you can throw diseased plants in the trash. We're loath to send organic matter to landfills, so we hope you'll only use this option if you absolutely have to.

or cardboard. A wide range of pests including aphids, thrips, whitefly, fungus gnats, leaf miners, and beetles can be attracted to the traps. Place traps on small stakes every 4 feet in the garden to help control flying pests. Sticky traps don't usually completely control a pest population, but they are helpful in reducing numbers and identifying problems early on.

ORGANIC PESTICIDES & FUNGICIDES

We intentionally left organic pesticides until the end of this chapter because we think of them as the last resort in a pest management strategy. This is not to say that these products should never be used—only that it is important to utilize all of the aforementioned techniques first in an effort to reduce the need for these products.

Pests and diseases can develop a resistance to even the most benign organic pesticide. If a spray kills 99 percent of the existing population, the crop may be saved in the short term, but the remaining 1 percent that were naturally resilient against the spray will survive to breed the next

generation of pests. These future generations will all have a better likelihood of genetic resistance to our controls.

Resistance to a pesticide doesn't happen overnight. The best way to avoid it is to use preventive strategies first, and not to rely on a single organic pesticide. By varying your control method, you prevent pests from selecting themselves for resistance to a specific method. Judiciously used, organic pesticides and fungicides are an important tool for maintaining a productive garden.

It's equally important to keep in mind that many broad-spectrum organic sprays can negatively affect the beneficial insects in your garden. Some pesticides affect only a narrow range of

BT RESISTANCE IN GMO FIELDS

In industrial agricultural operations, Bt genes have been inserted into genetically modified crops to create plants that produce their own insecticides. In some areas, the pest populations quickly developed resistance to the pesticide, reminding us that these types of controls need to be used judiciously and appropriately.

insects, but even these can kill innocent bystanders. For example, Bt (*Bacillus thuringiensis*) is a beneficial bacteria spray intended to manage caterpillar pests like cabbage loopers and tomato hornworm; however, it will also kill any butterfly larvae in the sprayed area. Many broad-spectrum sprays can damage or kill a huge range of insects, including pests, beneficials, and innocent bystanders.

When pest issues become severe, organic sprays like spinosad can help save crops.

In addition to the sprays listed below, there are many other organic pest and disease sprays available online and in garden stores, and there are always new ones being released. Research any spray thoroughly before use to learn how to safely and responsibly apply it.

Soap spray. Probably the most commonly used home gardening spray, a simple mixture of soap and water can help control or eliminate hordes of insects. It's especially useful for aphids, but it works on any soft-bodied insect—including spider mites. This spray is nontoxic and safe to use around children and pets. The soap coats the insects and breaks down their cell membranes, drying them out and killing them relatively quickly.

Soap spray should be used judiciously on crops, because heavy and consistent applications can result in phytotoxicity (damage to the plant). Spraying early in the morning will help minimize this. To control aphids, spray the soap every 2 to 3 days for a week or two and then stop applications. These repeated applications in rapid succession should disrupt the reproductive cycle of the insects.

To make your own soap spray, mix 1 tablespoon soap (liquid dish soap, castile soap, or a specialized insecticidal soap) with a quart of water in a spray bottle. Shake vigorously and spray it on insect populations. This spray should keep indefinitely.

Bt (*Bacillus thuringiensis*). This is a bacteria that, when ingested by caterpillars, prevents the insects from eating, killing them by starvation. Different strains of Bt affect different species of caterpillars, and only some strains are approved for use in organic agriculture.

Bt is typically sold in liquid form, which is then diluted and sprayed directly onto the leaves of affected crops. Bt remains active and alive only when wet, so there is only a short period of time when it can harm nontarget insects. Apply Bt when you have identified a serious infestation of a caterpillar pest and spray as needed to keep populations under control. Consider alternating spray applications with spinosad (described below) to slow down resistance-building in your pests.

Spinosad. Similar to Bt, spinosad is derived from a naturally occurring soilborne bacteria (*Saccharopolyspora spinosa*). A compound found in the bacteria is isolated and manufactured into a liquid organic pesticide. The compound works by overstimulating the nervous system of target insects including caterpillars, thrips, mites, beetles, and borers. The spray is not persistent in the environment but can affect a wide range of insects, so must be used carefully. Avoid spraying on flowering and fruiting plants to prevent damage to beneficial insects.

Neem oil. A potent broad-spectrum fungicide and pesticide, neem oil is derived from an Asian evergreen tree (*Azadirachta indica*), also called neem tree or Indian lilac. The seeds and fruits of the tree are pressed to release this strong oil, which deters an incredible range of pests (aphids, beetles, cabbageworms, caterpillars, fungus gnats, leaf miners, mites, scale, thrips, and whiteflies) and diseases (anthracnose, black spot, powdery mildew, and rust).

It can be harmful to beneficial insects as well, so use only when necessary and avoid applications when pollinating insects are visiting your plants. Apply once and watch to see results; often one to three applications is adequate to control a problem.

Copper sprays. Copper is a potent natural fungicide that controls many fungal and bacterial crop diseases. It's widely used by commercial organic growers to slow the spread of early blight in tomatoes in the Midwest and on the East Coast. There are many copper-based sprays and powders available to the home gardener, some of which are certified organic and some of which aren't. Copper works best as a preventive spray

LABELS CAN BE MISLEADING

Read labels of all materials carefully before buying. Chemical pesticides may be packaged in very natural-looking imagery. The safest way to identify appropriate materials is to look for an organic certification logo.

or when a disease is first showing up, so have it on hand and be prepared to spray it if you have a disease problem in your garden that necessitates its use. Although it is a naturally occurring metal, copper is toxic at high concentrations, so it's absolutely critical that you mix it properly (or use a premade spray) to prevent damage to your plants. It can also build up in your soil if used excessively over time. We recommend using copper sprays only when you have no other option to control a disease.

Iron phosphate. An easy-to-apply slug and snail deterrent, iron phosphate is a naturally occurring soil component that has been isolated as an organic pesticide. It is known to be nontoxic to humans, pets, and wildlife and replaces an older form of slug bait known as metaldehyde. Metaldehyde is still available as a pesticide, but is highly toxic and should be avoided. Read slug bait labels carefully.

Iron phosphate typically comes in pellets, which are sprinkled on the surface of the soil across the bed or surrounding sensitive crops. Iron phosphate can be expensive and will mold in wet conditions, so we recommend using it judiciously through the season. It's most effective when applied to protect germinating and newly transplanted crops when they are small enough to be fatally damaged by slugs or snails. Once crops are larger, reduce or eliminate use.

MANAGEMENT STRATEGIES
FOR PESTS & DISEASES

PEST OR DISEASE	MANAGEMENT STRATEGY
INSECTS	
Aphids	Selective plant removal, sticky traps, attracting beneficial insects, soap spray, neem oil
Beetles	Floating row cover, selective plant removal, neem oil
Cabbage loopers	Floating row cover, Bt, spinosad, attracting beneficial insects
Carrot rust flies	Floating row cover, nematodes
Corn borers	Attracting beneficial insects, selective plant removal
Corn earworms	Attracting beneficial insects, hand picking, Bt
Cutworms	Hand picking, physical exclusion, nematodes
Hornworms	Floating row cover, Bt, spinosad, attracting beneficial insects
Imported cabbageworms	Floating row cover, Bt, spinosad, attracting beneficial insects, neem oil
Leaf hoppers	Floating row cover, nematodes
Leaf miners	Floating row cover, nematodes
Root maggots	Floating row cover, nematodes
Slugs	Iron phosphate, diatomaceous earth
Snails	Iron phosphate, diatomaceous earth
Spider mites	Attracting beneficial insects, selective plant removal, selective leaf and stem removal, neem oil
Squash borers	Floating row cover, selective plant removal, delayed plantings, traps
Stink bugs	Floating row cover, diatomaceous earth, soap spray, neem oil
Thrips	Attracting beneficial insects, floating row cover, sticky traps, neem oil
Whiteflies	Attracting beneficial insects, floating row cover, sticky traps, neem oil
Wireworms	Cultivation, crop rotation

PEST OR DISEASE	MANAGEMENT STRATEGY

ANIMALS

Pest	Management Strategy
Birds	Floating row cover, bird netting, hot pepper spray
Cats	Fencing, hot pepper spray
Dogs	Fencing
Deer	Fencing
Groundhogs	Fencing
Rabbits	Fencing
Rats	Traps
Raccoons	Fencing, bird netting, floating row cover
Voles	Traps

DISEASES

Disease	Management Strategy
Anthracnose	Select resistant varieties, selective leaf removal, selective plant removal, copper spray, neem oil
Bacterial wilt	Select resistant varieties
Blossom-end rot	Water consistently, adjust soil nutrient balance
Clubroot	Increase soil pH, crop rotation
Damping-off	Provide air circulation, reduce watering schedule, reduce fertilization
Downy mildew	Selective plant removal, selective leaf and stem removal, baking soda spray, copper spray
Early blight	Selective plant removal, selective leaf and stem removal, copper spray
Fusarium wilt and rot (crown rot)	Select resistant varieties, selective plant removal, selective leaf and stem removal, copper spray
Gray mold	Selective plant removal, selective leaf and stem removal, baking soda spray, copper spray
Late blight	Selective plant removal
Leaf spot (cercospora)	Selective plant removal, crop rotation, baking soda spray, copper spray
Leaf spot (septoria)	Selective plant removal, crop rotation, baking soda spray, copper spray
Mosaic virus	Select resistant varieties
Powdery mildew	Selective plant removal, selective leaf and stem removal, baking soda spray, copper spray, neem oil
Rust	Selective plant removal, crop rotation, neem oil
Scab	Select resistant varieties, crop rotation
Verticilium wilt	Selective plant removal, crop rotation

PART

4

CREATE EFFICIENT SYSTEMS

↓

PLANT MORE OF THE CROPS YOU WANT & WATER THEM EFFICIENTLY.

The systems you use in your garden play a huge role in expanding your harvest while using your time efficiently. During the growing season, if you don't have tools and transplants ready when you need them, or your crops get water stressed because your irrigation system isn't dialed in, you've missed an opportunity to get more out of your garden. Skilled growers invest a lot of effort to set up their systems, and they refine them each year. This pays dividends in higher yields and less time addressing problems.

CHAPTER ELEVEN

SETTING UP A HOME NURSERY

If you have the time and resources, we encourage you to produce as many of your own transplants as possible, whether from seed or cuttings. Transplants, also known as "starts," are crops grown in a controlled setting prior to establishment in the garden. By establishing your own nursery—simply a protected space with adequate light and warmth to start plants indoors—you'll be able to care for thousands of small plants in very little space. Growing your own transplants lets you choose varieties that taste great and produce best in your climate, and ensures you have the right number of plants available at the right time.

In addition to getting plants off to an early start in spring, a nursery also allows you to start successions of crops through the season, when space is not available in the garden beds. This enables you to have good-size transplants waiting and ready to plant as soon as space opens up. Even in the peak of summer, you may be growing transplants of fall brassicas and late plantings of summer squash to replace early-season crops in your beds.

inside a home nursery

Creating and managing a nursery can be one of the most enjoyable aspects of food production. The controlled growing conditions ("controlled" being a relative term) in a nursery setting allow you to monitor and tend to your crops more closely. This ensures early success and allows you to select the most robust and healthy plants for graduation to the garden beds.

In order for your nursery to be successful, it will need light, heat, water, and ventilation. You can count on Mother Nature to take care of those elements in your garden beds (at least most of the time), but in your nursery, the environmental conditions are dependent on what you provide.

Adequate lighting. An ideal in-home or greenhouse propagation area should receive a full day of direct sun exposure. Areas like this might be hard to come by, but any space that receives a few hours of sunlight will make growing transplants easier. If the propagation area gets less than 12 hours of sunlight, your plants will be healthier with supplemental lighting.

Heating. The primary reason you're growing transplants in a nursery is that outside temperatures are too low for the plants to survive (or at least to grow properly). Therefore, it's essential that your propagation space be kept warm. Keep in mind that heated areas may also require more frequent watering, as heaters will inevitably reduce the humidity of the room.

Water. Make sure you have easy access to a water source, such as a hose spigot or sink, for irrigating your transplants and washing out containers for reuse.

Humidity. If you live in a cold climate and your propagation area is inside a building/house with forced-air heating, you might need to provide additional humidity to your propagation area. This can be done with a humidifier, with regular misting of plants, or by placing a shallow tray filled with a layer of gravel or perlite and water in the propagation area. Keep the tray filled with water three-quarters of the way up the gravel.

Potting up tomatoes from a 72-cell plug tray to 4-inch pots

THE HOME NURSERY

Wherever you build your nursery, you should allow room for certain key elements. In addition to the following, you may want to set aside some space outside of the propagation area for hardening off flats of plants that are ready to be transplanted, or for holding them if you don't yet have a space open for planting. Consider a semishaded area where containers won't dry out as quickly and where plant growth can slow down to prevent the starts from outgrowing their pots before planting time.

1. POTTING BENCH. This might be inside your propagation space or in a separate area that can handle more debris and moisture. You'll use this space to mix and screen soil and fill your flats and pots. A potting bench can even be a covered space outdoors where cleanup isn't as important, or an enclosed space that's easy to wipe down and sweep up.

2. PROPAGATION AREA. This is the space where flats and pots of newly sown crops will live while they're germinating and growing for the first several days, weeks, or months of their lives.

3. STORAGE. You'll need some space—perhaps in the propagation room, under the potting bench, or in a tool shed—to store soil mixes, tools, and empty flats and pots.

4. VENTILATION. A fan improves air circulation and helps minimize disease issues.

This is only necessary in rare circumstances; regular watering of transplants is usually sufficient to maintain adequate humidity.

Ventilation. Every growing area requires adequate ventilation to minimize disease problems. Consider the location of windows, vents, and doors when locating your propagation area. Air movement is essential to help control pest and disease issues, so placing the propagation space in a room with windows or doors will allow for a low-tech solution. Ideally, you'll be able to vent on opposite sides of the room for cross ventilation.

Depending on weather conditions, a window can remain cracked open for part of the day or the door can be propped open even for a short time each day, allowing new air to move into the room. In an area with poor ventilation, a small fan can be used to keep air circulating.

Convenience. You want to make it as easy as possible to check up on your little plants. Placing the nursery in a location that is visible and frequently visited will make it a lot easier to check moisture levels, add fertilizer, and take note of germination rates.

locating the nursery

You may have a few options when considering where to place your nursery. Your choice depends on your space, budget, and the quantity of starts you plan to grow. In most cases, you'll be choosing between growing your starts inside your home, in an attached sunroom, or in a separate greenhouse.

IN-HOME PROPAGATION

An in-home propagation area can be inexpensive and easy to set up, yet highly productive. In fact, you can grow hundreds of transplants in a few square feet of space, and flats can even be stacked vertically with the right kind of setup.

If you're planning to grow your transplants inside your home, remember that natural light will be very limited. Even if they're near a south-facing window, indoor propagation spaces will require artificial lighting. Also keep in mind that a propagation area is necessarily a little messy. Consider placing a protective waterproof covering on top of any carpet, hardwood flooring, or other delicate material you may be working over. It also pays to clean the area regularly to avoid buildup of dirt and mold.

The invasion of insect pests is a real possibility when keeping transplants in the home. A well-tended, well-ventilated growing area can operate pest-free, but even the best-managed space can succumb to the will of nature. In the early years of our business, we grew transplants inside our house, just outside of our roommate's bedroom door. It worked great until he started complaining about the cloud of fungus gnats he woke up to every morning. See Management Strategies for Pests & Diseases (page 162) for ideas on how to prevent and recover from insect invasions.

SUNROOMS & ATRIUMS

If you're lucky enough to have a sunroom or atrium attached to your home, this can provide an ideal location for transplant propagation. The best atrium growing areas are on the east, south, or west side of the house and are not heavily shaded by trees or nearby buildings. A very sunny, warm sunroom may be able to support transplants with little or no supplemental lighting.

If your sunroom is attached to the exterior of the house and does not receive supplemental heat from the house, keep a thermometer in the room to monitor temperatures and supply additional heating as necessary.

STAND-ALONE GREENHOUSE

For most gardeners, the idea of setting up a stand-alone greenhouse is very tempting. Greenhouses can provide great winter work spaces and are amazing spaces for starting transplants and producing crops all year long. The only real drawbacks of constructing and managing a greenhouse are the cost and relatively large amount of space they require.

Greenhouses can be expensive to purchase, expensive to build, and expensive to operate. That being said, they are one of the best tools you can invest in to increase the overall yield of your home garden space. You can use a small portion of your greenhouse for propagation, and the rest for the production of heat-loving crops in summer and hardy crops in winter. (See page 235 for more about greenhouses.)

outfitting your nursery

↓ You won't need many supplies on hand for a successful nursery, but the right materials are essential. Here's a look at what you'll need to create the right environment in your propagation area.

LIGHTING

In indoor and partially shaded spaces, there is rarely enough natural light to keep your propagation area in business. Supplemental lighting will ensure that your seedlings grow quickly and stay healthy. Light-stressed plants will develop thin, weak, "leggy" stems and will be more susceptible to pest and disease attacks and, ultimately, poor productivity. Setting up a working grow light doesn't require an electrician's license and doesn't have to cost an arm and a leg, but high-tech options are available for those who love their lumens.

Fluorescent lights on adjustable rope clip hangers are easy to raise and lower as the crops in your nursery change.

TIME

Most vegetable transplants will grow at a healthy rate when provided with 14 to 16 hours of light per day. If your growing space has little or no natural light, lights can be manually turned on and off or set on a timer to run for 14 to 16 hours per day. To mimic natural light cycles, turn on your lights early in the morning and run them until early evening. You may be able to turn your lights off for portions of the day that the nursery receives strong natural light.

INTENSITY & SPECTRUM

The intensity (or lumen output) of growing lights varies widely. The lower the intensity of the bulb, the closer it should be to the plants. Most grow lights will come with large reflectors, which help direct all of the light waves toward the plants, using the available intensity more efficiently.

Blue is good. Pay attention to where your lights fall on the color spectrum. Generally speaking, light at the blue end of the spectrum supports vegetative growth, while light at the red end of the spectrum supports flowering and fruiting. Keep in mind that plants do need a little bit of red light for vegetative growth, so lights that provide a true full spectrum of colors are best for transplant production.

If you're keeping plants under lights through maturity, you may need several types of lighting to provide light at various points in the spectrum. However, if you're producing young plants for outdoor production, general full-spectrum lights (metal halide, fluorescent, or full-spectrum LEDs) work perfectly fine.

Watch those Kelvins. The color of light from a bulb can also be expressed in Kelvins. This can be confusing because the term "Kelvin" is typically used to indicate temperatures. However, with light bulbs, it refers to the frequency of the light wave, which corresponds to the color of the light. To help vegetable plants produce healthy transplants, the bulbs should have a Kelvin rating of 5000 to 6500K (again, the blue end of the spectrum). If you plan to grow crops to maturity indoors, and need to support flowering and fruiting, the bulbs should have a Kelvin rating around 2700K (the red end of the spectrum).

FLUORESCENT VS. LED VS. HID

Your main options for supplemental lighting are fluorescent, LED (light-emitting diode), and HID (high-intensity discharge) bulbs. HIDs require large amounts of electricity, making them expensive to operate. Fluorescent bulbs and LEDs have much lower energy requirements, but also provide a lower quality of light.

Fluorescent. These are the least-expensive type of fixtures you can use for your lighting setup, and they use a very small amount of energy to operate, making them very cost effective. Make sure to choose bulbs with a Kelvin rating of 5000 to 6500K. Remember that fluorescent bulbs contain very toxic chemicals (including mercury), so take care not to break them, especially when managing a space in the home.

Because of their low lumen output, fluorescents must hang 1 to 3 inches above the tops of your plants in order to provide sufficient lighting. Most growers will install these lights on chains or ropes, so that the lights can be adjusted as plants grow. Another option is to place plants on shelves that can move up or down. This height requirement also necessitates having different fixtures for plants that are at different stages of growth. You can't provide proper lighting to 1-day-old seedlings and 3-week-old seedlings from a single fluorescent fixture.

Grow lights and stacked shelves allow you to propagate a large number of plants in a very small indoor space.

Fluorescents are great for small production areas and for lighting stacked shelves of propagation flats. Consider using or building a series of shelves, each with its own light fixture. The best fluorescent bulbs for transplant production are T8. A good fixture will have four parallel 4-foot-long T8 bulbs, which fit perfectly over two flats of plants.

LED (light-emitting diode). These bulbs are relative newcomers in the world of plant production. They are a great choice for a home nursery because they consume a tiny amount of energy and don't generate much heat, so are an energy-efficient way to provide supplemental light to your starts. Initial cost is typically higher than for

We all know that electricity and water don't mix, so be very careful when using electric appliances during propagation. Here are a few tips to help you stay safe.

× Do *not* use an electric blanket or other heating element that isn't designed for propagation.

× Keep plugs and extension cords away from the ground and from sources of water (we like to hang them from the ceiling).

× Shut off power and move equipment as necessary when watering.

× If you'll be supplying electricity to a greenhouse on an ongoing basis, hire a licensed electrician to run a separate circuit with GFCI outlets and moisture-proof covers.

× Only use an extension cord run from the house to the greenhouse for temporary, short-term use. Be sure it's plugged into an outlet in a dry location (or make sure it has a moisture-proof cover). A GFCI outlet is the safest option for this.

× Don't exceed the recommended load for extension cords (heaters and metal halide lights use a lot of electricity).

× Use extension cords that are rated for outdoor use.

LEDs do not distribute light for great distances, so fixtures should be kept relatively close to the tops of your plants. The exact placement of LEDs can vary depending on the intensity of the light and the brand. Most are placed between 12 and 30 inches above the plants. LEDs do not emit heat, so if you notice any legginess in your transplants, lower the fixture closer to the tops of the plants. The limited light distribution means that several fixtures may be required to cover your propagation area. In fact, the biggest risk with LEDs is that you may accidentally invest in a set of fixtures that don't provide enough light coverage for your plants.

HID (high-intensity discharge). These lights are considerably more expensive than fluorescents or LEDs but can cover a larger area with a single bulb. The ballast (power source) and light fixture are typically separate units in an HID setup. Many professional growers will use HIDs to encourage robust growth and high yields when growing crops to maturity in an indoor setting. Due to their high intensity, HID bulbs can stay in place and be fixed several feet above your transplants. They generate a lot of heat and can actually burn plants if they're placed too close to the foliage.

There are two principal types of HID lighting. Metal halide bulbs are best for vegetative growth, and if you're growing transplants for planting in an outdoor growing location, they should be the only light type you need. With a large-enough table space and reflector, a single metal halide light fixture should be sufficient for all of the transplant production needs of a home-scale grower. High-pressure sodium bulbs are best for flowering and fruiting of crops. They're the bulbs to use if you plan on keeping plants in an indoor setting through their entire life and into fruit production. We encourage serious cost/benefit calculations before investing in the materials and energy to provide season-long supplemental lighting to your vegetables.

fluorescents, but LEDs last longer and use less energy so they may pay for themselves over time. LEDs also don't contain toxic mercury. As LED technology continues to improve, their cost will continue to drop.

LEDs are available in a vast array of colors marketed for specific periods of the growing process (vegetative, flowering, etc.). For vegetable transplant production, look for a light that contains as many colors of the spectrum as possible, but primarily has the blue and red spectrums in a ratio of 5:1.

HEATING WITHOUT ELECTRICITY

If you'd like to avoid a spike in your utility bills, there are ways to heat a greenhouse without electricity or major infrastructure investment. One effective (though admittedly labor-intensive) option is to build a compost bin in the greenhouse. Start a bin with fresh compost early in the spring, and then place your seedling flats on a wire screen on top of the bin. The heat from the compost will warm the flats. Setting up low hoops and row cover over the plants will help maintain warmer temperatures at night. You'll need a significant amount of compost to generate enough heat to make this system work: A bin 3 feet by 3 feet by 3 feet or larger should be adequate.

If you have livestock such as chickens or goats, you can bed them down at night in the greenhouse to generate additional warmth for germinating seeds. Use appropriate fencing to make sure the animals don't eat your tasty young seedlings. We don't recommend housing animals in production greenhouses while

In this greenhouse, black drums of water absorb heat from the sun during the day and radiate it out at night.

crops are growing to avoid contamination of produce.

A wood stove can work well to heat your greenhouse, if you have a supply of wood and the time and skill to properly manage it. You'll also need to carefully monitor the humidity of the greenhouse.

Rocks, bricks, or jugs of water painted black and placed in the propagation area will absorb heat during the day and radiate it out at night. We've found that "heat sinks" like these are helpful, but are not sufficient on their own to supply enough heat for effective germination.

PROVIDING ADDITIONAL HEAT

If your propagation area is inside a house or building, you may not need an additional heat source if the ambient temperature is appropriate for seed germination (see the Planting Dates Worksheet on page 274 for germination temperatures for different crops). If you're germinating seeds that don't need particularly high temperatures, you may be able to get away with using an unheated cold frame or low tunnel (see page 232), depending on your climate and the time of year.

Many propagation areas, however, do require supplemental heat. To supply heat, you have two options: Heat just the area where you'll be germinating seeds or heat the entire nursery.

A heat mat with a row cover over it is an efficient way to heat a small germination area.

HEAT MATS

If the nursery space is just a little too cold, consider purchasing a heat mat designed for propagation. These are available in various sizes, and are perfect for warming up a small space. They also work well in an outdoor greenhouse if you have access to electricity. To provide additional warmth at night, you can cover the flats with a cold frame or row cover. Since a heating mat will only heat a limited volume of soil, this option works best for shallow seeding flats.

HEAT CABLES

If you need more heated space than a heating mat can provide, you can set up your own heated propagation table using heat cables and a thermostat. Follow the instructions included with the product closely when setting up to avoid risk of shock and to ensure effective and even heating. This structure can also be covered at night with a cold frame or low tunnel to provide extra protection.

SPACE HEATER

Another option is to set up a space heater underneath the table or rack that your transplants will be resting on, covering the table with a row cover or greenhouse plastic at night. Make sure there's enough space between the heater and the table (and anything else) to prevent fire danger. Keep in mind that the heat will rise; using a small fan will improve horizontal distribution of the heat. Remember to unplug and remove the heater when watering.

HEATING A GREENHOUSE

Heating an entire greenhouse is usually done with a propane or LP gas heater hung from the framing of the greenhouse. Setting up the necessary infrastructure for heating an entire greenhouse is best done by a professional; you can consult with the company you purchased the greenhouse from for more information (see Resources, page 292, for suppliers). Some professional growers are also experimenting with alternative fuels, such as waste vegetable oil, waste motor oil, and biodiesel for heating their greenhouses.

For most home-scale production growers, heating an entire greenhouse for propagating a small number of starts is an unnecessary monetary and environmental expense. We suggest setting up a heated area inside a production tunnel, or constructing a small, separate kit greenhouse for propagation.

GROWING MEDIA

Store-bought growing media are often referred to as "soilless" mixes. These mixes are typically composed of ingredients such as peat moss, vermiculite, perlite, shredded coconut husks (coir), and compost.

These mixes are devoid of garden soils because soils do not generally provide the aeration, drainage, and water-holding capacity necessary to maintain plant health in the controlled environment and containers of a nursery. Growing media are designed to hold water while still providing aeration and drainage in the artificial environment of a plant pot where insects and microbes are less present.

You can find blends made specifically for germination, propagation, or just general use. Soilless mixes are often sterilized to eliminate disease spores and kill weed seeds. In a home nursery setting, sterilized soil mixes are not absolutely necessary and, in fact, the lack of soil microbiology can lead to plant health issues during grow out. Growers will have different opinions about the cost/benefit analysis of sterile mix based on their personal experiences. We prefer to use a nonsterile mix with compost.

GERMINATION MIX

It's best to grow transplants from seed in a special soil mixture known as germination mix. A germination mix is a finely screened, lightweight planting medium. The soil's small particles make it easy for young seedlings to push up through the mixture without being damaged or trapped under a large piece of bark or other material you might find in a standard potting mix.

Depending on your scale of production, available time, and budget, you can buy ready-made germination soil or mix your own. A standard germination mix consists of peat (or coconut coir), perlite, and lime. These mixes can be relatively expensive, but a large bag can last a long time for the home-scale nursery. If you'd like to create a germination mix at home, try our recipe at right.

POTTING MIX

As your plants mature, they'll require more nutrients than a germination mix can provide. So, if you pot up your young transplants to larger containers, it's important to use a soil mixture with higher nutrient levels and more active microbiology.

MAKE YOUR OWN MIX

Creating your own soil mix can be a fun and cost-saving activity. Keep in mind, though, that buying bulk materials to create mixes will take up a considerable amount of space, and properly mixing soils requires time and patience.

GERMINATION MIX

× 15 parts peat

× 4 parts compost

× 4 parts perlite

× 0.25 part dolomitic lime

× 0.25 part fine, granulated, balanced, organic fertilizer

Mix the ingredients thoroughly. We like to use 5- and 1-gallon buckets to measure out the ingredients. You can mix up a batch in a wheelbarrow by combining 2 cubic feet of peat (about three 5-gallon buckets), ½ cubic foot of compost (a little less than one 5-gallon bucket), ½ cubic foot of perlite (a little less than one 5-gallon bucket), 4 cups of dolomitic lime (¼ gallon), and 4 cups of fertilizer (¼ gallon). You can keep this mix in a clean plastic or metal garbage can with a lid for use all spring. It needs to stay dry to prevent the fertilizer from decomposing. Moisten only the amount of mix you'll be using, just before seeding.

POTTING SOIL MIX

× 1 part compost

× 1 part sand

× 0.5 part peat or perlite

Mix the ingredients thoroughly and use right away or store out of the elements for as long as necessary.

Broccoli growing in 32-cell plug trays

When potting up, you can use any store-bought, organic, all-purpose potting soil. Most potting soils contain perlite, bark, compost, and a variety of other materials (like worm castings and manure) and should be a good starting point for almost any vegetable transplant. Because nursery plants require frequent watering (and thus leach nutrients), even the most nutrient-rich potting mix may require regular doses of liquid fertilizer as the plant grows. If you'd like to mix your own potting soil, try our recipe on page 175.

CONTAINERS FOR EFFICIENT SEEDING

Seeding "flats" or plug trays are a great option for producing a large number of plants efficiently in a small space. Generally, flats are always the same dimension: approximately 10 by 20 inches (but with the lip, they are actually about 11 by 22 inches). Flats are built to hold virtually any number of plants—open flats can be seeded with hundreds of plants. Plug trays contain a set number of individual cells, which can help ease the transplanting process. Another option is to use soil blocks, which can be very useful in certain situations. Or you can simply start your plants in egg cartons, yogurt cups, or other containers you have on hand.

OPEN FLATS

You can purchase open flats with no internal spatial divisions that are designed to be filled with growing media and seeded into. These flats are inexpensive, easy to use, and can be utilized for starting a variety of plant types. When using open flats, you must be very careful when removing young seedlings for potting up or transplanting. Roots will have grown together, and it is very easy to damage or kill plants when removing them from these trays. Some growers use these flats exclusively, but we typically avoid seeding into shallow open flats because of the difficulty of thinning and transplanting after germination.

PLUG TRAYS

Most of our transplants are seeded into plug trays. Plug trays can be purchased in a wide array of cell sizes and numbers of cells per tray. We primarily use 72- and 128-cell trays. Plug trays allow you to use your seed efficiently. We typically place two seeds per cell to reduce future thinning needs but ensure good germination rates and effective use of space. Seeds with very high germination rates or seeds that are very expensive may be sown one per cell to stretch the seeds as far as possible and eliminate the need for thinning.

Sow 20 percent more. Keep in mind that, even under ideal conditions, seeds will rarely have 100 percent germination rates. Unless they're seeded with several seeds per cell, most plug trays will have some empty cells. For example, if you want to plant 72 peppers in the garden, make sure to sow several seeds per cell in a 72-cell tray or to sow more than one tray of the crop. As a general rule, always seed at least 20 percent more than you plan to use in the garden, to account for unhealthy seedlings, mortality, and other unpredictable elements.

Transplanting. Most crops can be effectively grown out in a 72-cell tray and held for a few weeks until transplanting. Crops held too long in plug trays can become rootbound or suffer from a lack of available nutrients, so keep an eye on growth and add consistent low doses of liquid fertilizer. Transplanting from plug trays is easy since the plant's roots will have filled out the soil in the cell by the time they're ready for planting. Most crops will lift out of the tray easily if gently tugged by the leaves. To assist with transplanting, turn the tray on its side, squeeze the bottom sides of the cell, and gently pull each plant out.

SOIL BLOCKS

Many small-scale farmers use soil blocks for transplant production. A soil block is essentially a cube of soil to germinate seeds in without the structure of a plug tray. A tool called a soil blocker is used to compress growing media into multiple small cubes, which can then be seeded with the crop of your choice.

Soil blocks are great because plant roots will stop growing when they reach the edge of the exposed block (roots don't like to be exposed directly to air). This prevents seedlings from becoming rootbound and can ease their transition into the garden, since root tips are ready to move beyond the edge of the block once they're placed into a garden bed. Soil blocks take a little more skill and care to use than plug flats, so be patient when learning the ropes of soil-block production. To ensure soil blocks hold together, use warm water when initially moistening the mix and make sure the mix is completely saturated. The consistency of the soil should be similar to a paste, much wetter than in other applications. This initial work will help the ingredients form strong bonds and maintain their shape.

OTHER CONTAINERS

Many growers, when producing transplants on a small scale, find it easy to locate free containers for seeding. Egg cartons are a time-tested seeding tray, as are yogurt cups. Pretty much any small disposable container will work. Make sure to poke holes in the bottoms of the containers you use, as drainage is essential to prevent root rot and to maintain healthy plants. The only real disadvantages of using random containers for propagation are the inefficient use of space and the increased risk of drainage issues.

LARGER CONTAINERS FOR POTTING UP & SOWING

Potting up seedlings into larger containers can be useful for growing cold-sensitive crops like tomatoes and peppers in the protective space of the nursery for a long period, or if weather conditions prevent you from planting outside for longer than

Making soil blocks: Make sure your soil mix is adequately moist, dunk your blocker in water periodically to help the blocks keep their shape, press firmly, and cover with a lightweight substance like peat moss after seeding.

anticipated. In general, though, because potting up requires additional time and materials, we encourage you to transplant directly into the garden from small containers whenever possible.

4-INCH POTS

These are typically used to pot up seedlings from plug trays, but it's possible to sow seeds directly into 4-inch pots, as well. Starting in 4-inch pots will eliminate the need for potting up crops like tomatoes and peppers, but will also necessitate more propagation space or a more limited selection of transplants. Cucurbits like squash and cucumbers are commonly seeded directly into 4-inch pots because they can become stunted if potted up or transplanted roughly.

Sowing seeds. If sowing directly into 4-inch pots, consider filling the pot almost entirely with potting mix and adding a thin layer of germination mix to the top of the container to sow into. The germination mix will create a good seedbed, and the underlying potting mix will provide the additional nutrients needed once the seedling's roots expand and the plant starts to grow.

GALLON-SIZE CONTAINERS

Depending on the type of crops you're producing and how you've scheduled your seeding, you may want to use gallon-size containers in your propagation nursery. If you like to start heat-loving crops like peppers and eggplant early in the season to get them as large as possible before planting out into the garden, potting up into a gallon-size container is probably a good idea.

Sowing seeds. Certain crops can be sowed directly into gallon-size containers. We often sow bulb onions and leeks into 1-gallon nursery pots. These large, deep containers allow them to grow out for several months before transplanting. Healthy onions can take a fair amount of handling during transplanting, so these can be seeded

Allium crops like onions and leeks can be easily sown into 4-inch pots, then separated and transplanted directly into the garden when ready.

thickly into containers and pried apart at planting time. Using larger containers reduces the need to worry about plants becoming rootbound.

PLANT TAGS

We can't stress enough the importance of proper and consistent plant labeling. Even if you're transcribing every activity into a notebook or have a photographic memory, it's incredibly easy to mix up identical-looking young transplants. With some crops, it's virtually impossible to distinguish among different varieties when the plants are young.

When seeding an entire flat with a single variety, we recommend putting at least two tags in each flat (in case one falls out). If seeding multiple varieties together in a flat, plan to place a tag in each row. Write the date, the type of crop, and the specific variety of the crop on the tag. With all that information, you may need to use both

Trays and pots can be storehouses for pests and diseases that can damage your crops. We recommend washing your pots with water and dish soap after each use, and then sanitizing with a mild bleach solution (1 part bleach to 10 parts water) or a peroxyacetic acid product. Pots should be allowed to air-dry and be stacked somewhere out of direct sunlight until the next use. If well taken care of, your containers should last for a number of years.

sides to keep things legible. Your tags can then be placed in the garden when transplanting to help you identify and differentiate between varieties when you're out in the field. Wooden plant tags written on with permanent marker are great because they usually last for a whole season, and can be left in the garden to decompose. Simple plastic plant tags written on with grease pencils can be much less expensive if you are growing on a large scale. These have the advantage of coming in bright colors, so they are easier to locate among large, healthy plants, but they must be collected and disposed of after each crop is cleared.

WATERING IMPLEMENTS

You'll want something that can provide a gentle stream of water for your transplants. A standard watering can or hose with a nozzle both work just fine; some growers even hold a watering can up to the end of a hose as a makeshift nozzle while they water. If using a nozzle, try to find one with a gentle shower setting for everyday use and a mist setting for moistening very small seeds and for misting grafted plants. If you build out a larger nursery, consider an automated overhead spray or bottom-watering system.

TABLES & RACKS

You'll need to place your nursery flats somewhere up off the ground while they germinate and grow out, to keep them away from ground-dwelling insects and small animal pests. Any kind of table will suffice for this purpose—even just a pallet or a piece of plywood set up on blocks—as long as it can get wet and dirty. Common greenhouse and nursery tables have tops built from metal grating so water drains easily away from the flats and so you don't have to worry about a warping and rotting wooden table. You can make your own nursery tables with lumber and wire fencing. Consider using the space below your tables for storing supplies that can handle getting wet, such as empty flats or closed containers of potting soil.

Go vertical with racks. When setting up a nursery in a small space, you can stack flats on a rack with multiple levels to take advantage of vertical space. Any storage rack will work well for this purpose if the space between shelves is adequate to grow out your plants. A minimum distance of 12 inches is best to give yourself the option of growing plants in this location for more than a few weeks. If you plan to pot up the crops and keep them in a stacked situation for more than a few weeks, try to use a rack with 24 inches or more between shelves.

If you're stacking trays in this way, it's important to make sure all levels get adequate light exposure. Attaching a fluorescent or LED light to the bottom of each shelf will create a very condensed propagation nursery.

Some growers stack flats directly on top of each other or use unlit space while the seeds are germinating. This does work, but the plants will need access to high-quality light the day they begin poking out from the soil. Failure to provide light quickly enough will cause the plants to become leggy and unhealthy, so be very attentive if you decide to stack propagation trays.

CHAPTER TWELVE

STARTING YOUR OWN TRANSPLANTS

Starting a nursery with a diverse collection of crops and varieties requires significant planning and a well-organized space. The actual time investment managing a home nursery is not incredibly significant; the key is to allow for short but frequent checkups and tasking. Setting aside a few minutes every day to check water needs, scout for problems, and seed new flats will pay huge dividends throughout the season. One key advantage of transplant production is that it gives you the ability to care for the plants in a controlled environment and get them off to a healthy start.

This 72-cell plug tray holds **14** different varieties of brassica crops.

scheduling your seed starting

↓ Refer to your planting plan (see page 39) to determine how many transplants you need to produce for each crop, and seed your flats accordingly. Again, be sure to plant at least 20 percent more than you'll actually need. For example, if you need 30 tomato plants, plan to seed 50 cells (two seeds in each cell). Thin to one plant per cell and transplant only the 30 healthiest tomatoes in the garden. If you're seeding 50 cells, you'll likely have only 35 to 45 germinate, leaving a little room for plants that just aren't healthy and for a few extra plants to give away/trade/sell at planting time.

As you're determining how many plants to start, plan your space carefully. You'll likely be starting seeds throughout the season, and

possibly potting up transplants, so be sure to leave enough space open for your propagation needs later in the season.

MAKE YOUR SCHEDULE

After you've determined the quantities you'll need for each crop, prepare your seed-starting schedule. You can use a single schedule for the entire garden that includes a sowing schedule for the nursery, as well as outdoor seeding and planting dates, or a separate calendar just for the nursery area.

Your seed-starting schedule should list the quantity of each crop to be sown in a given week. At our nursery, we plan to do each week's seeding

Create Efficient Systems

STARTING SEEDS: HOW THE PROS DO IT

Seed starting is a fairly simple process, but it's important to do it properly to ensure even, consistently high germination rates. Most seeds should have a germination rate of 80 to 95 percent when you purchase them, but it's up to you to provide the conditions necessary to achieve this.

1 When you're ready to sow, fill each flat or pot up to the top with germination mix or screened potting soil. Use the bottom of a similar-size container to gently tamp down the soil in your pot, creating a level seeding surface ½ to 1 inch below the rim of the container. If the soil mix is dry, moisten it thoroughly before sowing.

2 Sow the seeds directly on top of the prepared seed surface and dust over the seed with germination mix until achieving the desired depth. Seeds require different sowing depths. Every seed packet should indicate how deeply to sow the seed in the soil, but usually smaller seeds are planted more shallowly than larger seeds, and as a general rule, seeds should be planted twice as deep as their diameter.

3 Sow more than one seed per cell or pot. In a small nursery setting, space is at a premium and you don't want to dedicate heated propagation space to empty containers. Sow heavily enough to ensure a viable plant in each container, and then thin to the healthiest plant once they emerge. For most crops two to three seeds per container should be adequate; sowing any heavier than this can make thinning difficult and stressful for the little sprouts.

4 Seeds need good contact with the soil to absorb moisture during germination. Simply watering in the seeds after planting will provide adequate seed-to-soil contact. Keep newly sown flats consistently moist during the entire germinating process. If a seed dries out for even a few hours, young sprouts may desiccate and die.

Rather than uprooting, use scissors to thin plants that don't like root disturbance.

5 Once the plants have germinated and are large enough to distinguish, cull all but one plant in each cell. Culling can be done by uprooting extra plants or simply by snipping them off with scissors. The appropriate technique may be dictated by the crop and its size at culling time. If in doubt, pinch or snip off the extra plants to prevent disturbing the young root system of your keeper plant.

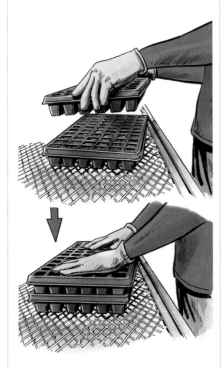

Use a similar-size container to gently tamp down soil to an appropriate depth for seeding.

on the same day each week (Monday), so the specific day of the year is included in the spreadsheet, to help keep us on schedule. If we make changes to the schedule and sow on a Tuesday or Wednesday, we note the change in the schedule. Additionally, if we adjust quantities (for example, if we planned to sow two trays of lettuce but only sowed one), these changes are also noted. This gives us an accurate record of when and how much we plant, and helps us project more accurately for future seasons.

We almost never follow our seed-starting schedule exactly as written. If a certain sowing of a crop fails, we need the flexibility to add to the next week's schedule to make up for it. Similarly, if early plantings of a crop are incredibly successful, we may choose to eliminate a sowing later in the season.

Depending on your location, your seeding schedule may last all year long, or it may stop for several months when the sunlight is not adequate for plant growth. Our schedule runs all year. Transplant production starts in January with onions and leeks and continues in earnest through September. During the months of October through December we don't start

SAMPLE SEED-STARTING SCHEDULE

DATE SOWN	CROP	VARIETY	PROPAGATION CONTAINER	AMOUNT SOWN	DATE GERMINATED
4/1	Peas	'Sugar Snap'	72-cell flat	2 flats	4/6
4/6	Scallions	'Nabechan'	72-cell flat	1 flat	4/13
4/6	Leeks	'King Richard'	4-inch pots	2 pots	4/14
5/1	Brussels sprouts	'Dagan'	72-cell flat	¼ flat (18 cells)	5/8
5/1	Brussels sprouts	'Hestia'	72-cell flat	¼ flat (18 cells)	5/8
5/1	Kale	'Lacinato'	72-cell flat	¼ flat (18 cells)	5/7
5/1	Kale	'Rainbow Lacinato'	72-cell flat	¼ flat (18 cells)	5/8

any seeds for outdoor transplanting, but we do sow flats of microgreens and sprouts for winter production.

To plan your seed starting for the season, you need to start by developing your planting plan and calendar (see Chapter 2). Once you've determined your crop amounts and transplanting dates, you can use the Planting Dates Worksheet (page 274) to figure out when to start your seeds. For example, let's say you want to have a broccoli start ready to transplant on June 1. You consult the Planting Dates Worksheet and note that broccoli takes about 4 weeks to grow from seeding

GROUP FAST-GROWING & SLOW-GROWING CROPS IN SEPARATE FLATS.

until it's ready for transplanting, so you'll want to start the broccoli seeds in your propagation area on May 1 (about 4 weeks before June 1). Some growers keep track of their seeding schedule right on their planting calendar, others make a separate seeding schedule to keep track of this information.

DATE TRANSPLANTED	NUMBER OF PLANTS TRANSPLANTED	CROP NOTES
4/12	105	Lost some plants when the edge of the seeding flat got too dry. Make sure to water twice a day on hot days.
4/25	65 (clusters)	Interplanted with spring brassicas. Still probably didn't need this many. Could plant half a flat next time.
5/1	30	Seeded approx. 20 seeds in each 4" pot.
5/20	10	Planted half a flat with Brussels and half with kale. Both germinated on the same day. Looking forward to seeing how these varieties perform. Planted each variety in a separate row to make it easier to keep track of them.
5/20	10	
5/20	15	
5/20	20	The rainbow type looked better, so I planted more of these.

GROUP CROPS TO MAKE TRANSPLANT PRODUCTION EFFICIENT

Group by rate of growth. We group fast-growing crops and slow-growing crops in separate flats. Since you're unlikely to ever need a full flat of a particular crop, this saves space and ensures that an entire flat is ready for transplanting at the same time. Here are some examples.

+ Fast-growing mixed big brassica flat (72-cell plug tray)—broccoli, cabbage, cauliflower, collards, kale

+ Fast-growing mixed head lettuce flat (128-cell plug tray)—butterhead, romaine, Bibb

+ Slow-growing greens (72-cell plug tray)—celery, celeriac, parsley, scallions

+ Slow-growing solanaceous flat (72-cell plug tray)—tomatoes, peppers, eggplant

Group by environmental needs. Grouping crops that need similar germination or grow-out temperatures can be helpful to ensure all of your sensitive crops get the conditions they need to be successful. For example:

+ Crops that need warmth—basil, eggplant, peppers, and tomatoes. These all need relatively high temperatures (80 to 90°F/26 to 32°C) for germination and grow out.

+ Crops that don't need warmth—lettuce and snap peas. These can germinate and grow in cool conditions. They might not be the crops to prioritize if you have limited warm space.

Group by seed size. Sow larger-seeded crops directly into larger containers to avoid potting up. For example:

+ Sow into mixed flats of 4-inch pots—cucumbers, melons, summer squash, winter squash

Plug trays can be filled and seeded quickly.

Using a hand seeder to add five to seven scallion seeds to each cell of a plug tray

seeding equipment

↓ As you scale up nursery production, hand seeding flats can start to take up a lot of time. Working quickly while still accurately seeding cells can present a challenge even to a seasoned grower. Fortunately there are a range of tools that can help you save time and increase seeding accuracy.

Hand seeder. This handy device is simply a small container with a single chute sticking out to the side. Fill the basin with your seed and gently tap the chute to help direct seeds into each cell of your flat or soil blocks. Using the tool effectively takes a little practice, but once you develop a feel for it, you can double your seeding speed and accuracy. Some growers prefer to use a pen or pencil to tap the side of the chute rather than their finger because they believe it gives them more control. If you like high tech, you can also find electronic versions of this tool that will do the seed shaking for you.

Wand seeder. Slightly more technical than the basic hand seeder, a wand seeder is a simple, small-scale vacuum seeder. They typically use a hand or foot pump to suck up and deposit seeds. These are especially useful for tiny, hard-to-handle seeds like lettuce.

Vacuum seeder. A vacuum seeder works by sucking seeds up to evenly spaced holes on a metal plate. Each plate is designed to fit a particular plug tray. For example, you can purchase a seeder plate for a 72- or 128-cell plug tray. Simply pour seed onto the tray and turn on the vacuum, and the seeds will be sucked up into a placeholder that aligns perfectly with each cell of your tray. Place the plate over the tray, release the vacuum, and presto, the whole tray is seeded with a single seed per cell. This is a relatively advanced and expensive piece of equipment, but if you're considering high-volume nursery propagation, it might be worth checking out.

Starting Your Own Transplants

propagating from cuttings

Growing transplants from seed is the most common form of propagation, but it's certainly not the only way to start new plants. Both annual and perennial crops can be grown from cuttings, and this is a quick and inexpensive way to expand your production.

ANNUAL CROPS

Growing annuals from cuttings will be most relevant if you're planning successions of long-season crops. Cuttings can be ready to plant more quickly than germinated crops; they can also make rare or expensive seeds stretch further in the garden.

GROWING BOTH ANNUAL & PERENNIAL CROPS FROM CUTTINGS IS A QUICK & INEXPENSIVE WAY TO EXPAND YOUR PRODUCTION.

Basil cuttings will quickly root out in a glass of water and can be used to make easy successions of your basil crop.

Herbaceous cuttings. These are taken from nonwoody, herbaceous plants like basil and tomatoes. They are the simplest type of cuttings to manage, because they are the most likely to survive and require the least oversight. They set new roots quickly, and many of them can be rooted out simply by placing them in a jar of water. (If you're planning successions of basil or tomatoes, rooting out cuttings in water is a great idea.)

PERENNIAL CROPS

You can take cuttings from virtually any perennial herb, including anise hyssop, lavender, mint, oregano, rosemary, and thyme, and grow them out in your nursery. Take cuttings in spring, summer, or early fall when plants have new vigorous growth. Don't take them when the plant is flowering, water stressed, or right before frosty weather.

Softwood cuttings. These are taken from soft, succulent, new growth of woody plants, including perennial herbs like rosemary and sage. New shoots and growth tips are cut when stems are springy and can be snapped off easily when bent.

Hardwood cuttings. These are taken from the mature branches of woody perennials, such as bay laurel and other large shrubs, when the plant is dormant.

LAYERING

Layering is the process of encouraging a plant's branches to contact the soil in the expectation that it will set out a new root system. Some plants like mint and rosemary will layer themselves

MAKING NEW PLANTS FROM CUTTINGS

The process below describes how to root cuttings in germination mix or perlite. However, many herbaceous and some softwood cuttings can be grown directly in a container of water (with no other medium). If employing this method, follow steps 1 and 5 below, but instead of placing the cuttings in germination mix, simply place them in a jar with water and make sure the leaves of the cuttings sit comfortably above the water line. Change the water every few days to reduce the potential of bacterial growth.

Dip ends of cuttings into rooting hormone.

Stick fresh cuttings into a flat of moistened perlite or germination mix.

Pot up rooted cuttings into containers filled with potting mix.

1   Use a pair of sharp, clean pruners or scissors to cut off a 3- to 6-inch piece of stem from the plant. Make sure you are cutting new, soft growth from healthy branches. Cut the stem at a 30- to 45-degree angle; the extra surface area this creates will enable the cutting to send out more roots. Take three or four cuttings for each plant you hope to propagate. Often cuttings will fail for no clear reason, so hedge your bets by making a few extra.

2   Dip the cut end of the cutting into rooting hormone powder. Rooting hormone powder is a condensed, dried form of a compound found in all plants. It's usually harvested from willow plants, which have an unusually high amount of the compound. Rooting hormone can be found at any garden center or online, and one package should allow you to make hundreds of cuttings.

3   Place the cuttings into a flat or small container filled with germination mix or straight perlite. Do not include compost or garden soil—you want to prevent the sensitive cutting from developing bacterial or fungal infections while it becomes established. Keep the cuttings in indirect sunlight while they establish root systems.

4   Keep the growing medium consistently moist. Cover the flat with a plastic cover (you can purchase one to fit over your flats, or you can make your own) and mist the cuttings with water as necessary to maintain a high level of humidity. Uncover periodically to allow evaporation of excess water and avoid fungal problems. Remove any dying or diseased cuttings to prevent the spread of pathogens.

5   When the cuttings begin to root, pot them up into a larger container. Allow them enough time to set out a root system that fills out the nursery pot, but not so long that they become rootbound. The easiest way to monitor their progress is to periodically lift one or two plants from the container (carefully) and examine the extent of root growth. Once they have grown healthy root systems, plant them into the garden. New plantings of herbs will take most easily when transplanted in spring or early summer.

freely in the garden, but you can encourage other plants to reproduce in this way, too.

The layering process is simple. If the branches are near the ground, simply bend them toward the ground enough to cover a portion of the stem with soil, then wait. If the branch needs extra help staying put, use a pin or landscape stake to help hold it down. Once the branch has rooted, simply use your pruners to remove the branch and find it a new home in the garden, or pot it up in the nursery to hold for future use.

ROOT CUTTINGS

Root cuttings are exactly what you might imagine: portions of a plant's root system that are separated and replanted to grow another plant. Root cuttings are especially easy to take from perennial plants that are harvested for their roots, such as sunchokes, horseradish, and echinacea.

When harvesting these crops, separate out as many healthy pieces of root as needed to supply your next crop and replant them in the garden at the appropriate spacing as soon as possible after digging them up.

DIVISION

Very similar to taking root cuttings, dividing plants is the practice of digging up the root mass of a plant and dividing it into multiple pieces. You can use a shovel or hand tool to separate the mass into a few distinct pieces. The number of divisions possible will depend on the size of the root mass. Make sure each piece has a healthy section of roots. Replant the divisions quickly and water them in well. This process works for rhubarb, chives, thyme, oregano, tarragon, and cardoons.

FOR SOME PERENNIALS, INCLUDING LEMON BALM, LAYERING IS A GREAT WAY TO PRODUCE YOUR OWN ROOTED CUTTINGS DIRECTLY FROM AN EXISTING PLANT.

Use a silicone clip to connect a fruiting variety to the rootstock.

grafting for increased productivity

A centuries-old technique for propagating all types of plants (most notably ornamentals, fruit trees, and vines), grafting can be used to improve your vegetable crop production. Long practiced in Asia and Europe for annual crops, grafting has begun to catch on in the United States as a way to boost yields, increase vigor, and reduce susceptibility to pests and diseases. It is commonly practiced on solanaceous and cucurbit crops.

The idea is to select a variety for the rootstock (the part of the grafted plant that will provide the roots and lower stem of the plant) that is hardy, vigorous, and productive, and a variety for the scion (the upper part of the grafted plant that actually produces the fruit) that meets your need for flavor, size, and shape. Grafting may be particularly helpful for heirloom varieties that taste great but have issues with disease or low productivity. Make sure to identify and order the appropriate seeds well in advance of the grafting season. Any available cultivar of a cucurbit or tomato crop should be feasible to graft.

To accommodate the slow growth rate of the seeds for the rootstock, they should be planted about 5 to 7 days before starting the seeds for the scion. Grafted plants need a few weeks to heal after the grafting process, so you might start them 2 to 3 weeks earlier than you would for nongrafted plants for the same crop (although grafted plants are typically more vigorous and often catch up to their nongrafted counterparts even if they're planted at the same time).

For solanaceous crops, it is key to order a specially selected rootstock variety seed from a reputable supplier. Challenging cucurbits such as cucumbers and melons are often grafted onto rootstocks of common, vigorous, hardy species such as gourds and squash.

GRAFTING NEW PLANTS

1 Sow rootstock seed 5 to 7 days earlier than the scion, since rootstock seed is typically slower to germinate. The temperature should be at least 65 to 75°F (18 to 24°C).

2 When plants have reached approximately 4 inches tall, with two sets of true leaves and a strong stem, they are ready for grafting. It is essential that the stem diameter of the scion plants and the rootstock plants are the same. Individual size of plants will inevitably vary slightly, so match each individual scion with a rootstock plant that has the most similar diameter. You will leave the rootstocks rooted in their containers and will bring the cut tops of the scion plants to the rootstock.

3 Wash your hands and clean all materials before handling the plants.

4 Using a new, clean, sharp razor blade or surgical scalpel, cut off the top of the rootstock plant at a downward 45-degree angle just below the cotyledons. Discard the top of the rootstock plant so it does not get confused for a scion.

5 Cut off the top of the scion plant at an upward 45-degree angle at a location that matches the diameter of the rootstock.

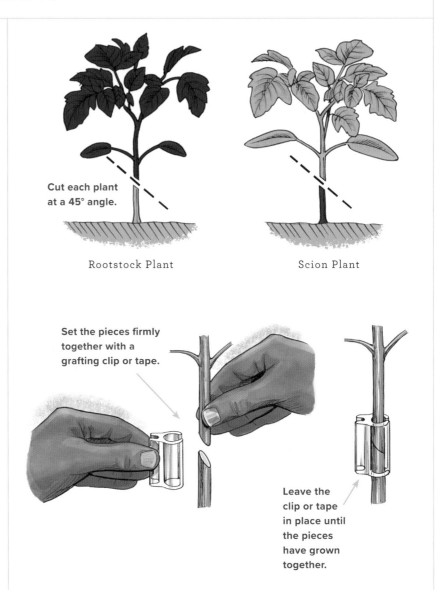

Cut each plant at a 45° angle.

Rootstock Plant

Scion Plant

Set the pieces firmly together with a grafting clip or tape.

Leave the clip or tape in place until the pieces have grown together.

6 Set the cut pieces together and firmly hold them in place while you attach a grafting clip or wrap the union in grafting tape or surgical tape.

7 As soon as possible, place the flat under a humidity tent and keep it in a dark place for approximately 1 week while the healing takes place. Keep flats at 65 to 80°F (18 to 26°C).

8 Check on the flat twice daily to make sure moisture remains on the side of the humidity tent. Spray with water as needed to maintain a high level of humidity. It is essential that grafted plants remain humid and warm through the healing process.

managing your nursery

↓ Managing a home nursery shouldn't take more than a few minutes a day. Changes happen quickly, though, so it's important to keep a watchful eye over the plants. Plan to check on the nursery a few times throughout the day to ensure that flats don't dry out and that the propagation area stays at the proper temperature.

WATERING

Depending on your scale of production, you can use a watering can, hose, or automated irrigation system to water your flats. If you're overhead watering, it's important that it can reliably deliver a *gentle* shower of water for your plants. Most seeds are planted very shallowly in the flats and germination mix is very light, so a heavy shower or stream of water can wreak havoc, washing seeds out of place or completely out of the flat.

Another technique is to bottom water your flats. This allows you to keep water off of the new growth and deliver the water to the root zone more efficiently. Thanks to the wonders of capillary action, if your flat is set in a water-holding container, water will wick up into the soil from below. You can set flats of plants in plastic tubs or on shallow plastic trays, or build a water-holding table. Add water to the reservoir around the flats as needed.

It can be tricky to bottom water flats before germination without overly drenching the soil or letting the surface dry out. Therefore, if you'd like to use this technique, it can be helpful to water overhead until seeds have emerged and then switch to bottom watering.

Because of their limited soil volume and shallowness, flats can dry out quickly. Plan to check moisture at least once a day, and potentially two or three times a day in hot, sunny weather.

When seeds are germinating, it's essential that soil remains evenly moist until the plants have sprouted and begun developing a root system. Once plants have sprouted, allowing the soil's surface to dry out a bit between waterings will reduce pest and disease issues. Be wary of over- and underwatering. The goal is to keep the

Overhead water with a gentle shower setting on a wand sprayer.

Bottom water into a tray that is slightly larger than your seedling flats.

This flat of summer squash in 4-inch pots is ready for planting.

planting mix damp but not soaking wet—like a wrung-out sponge. Generally speaking, it's best to water flats before noon so the plants' foliage can dry off before dark.

Mature cuttings and grafted plants can be managed in much the same way as plants grown from seed. Once cuttings are potted up out of their initial root-out period, they can be overhead or bottom watered.

FERTILIZING

Proper nutrition for young plants is crucial, but it can be one of the more challenging aspects of nursery management. When plants germinate, they have a small store of nutrients that enable them to establish themselves and begin searching for other nutrient supplies in the surrounding soil. Most germination mixes, including your own, should not contain high doses of granular fertilizers. Fertilizers contain a variety of salts, and high

levels of fertilizer can affect the ability of sensitive seeds to absorb water, reducing and potentially entirely preventing germination. It's better to allow plants to germinate and then begin adding small amounts of fertilizer on a regular basis.

Propagation containers are shallow and require frequent watering. Therefore, nutrients leach much more quickly than they do in garden soil. Given this, frequent low doses of soluble fertilizer are much more effective than less-frequent doses at full concentration. Many nursery growers water transplants with low doses of liquid fertilizer on a consistent basis. Depending on the concentration of nutrients, plants can be fertilized every day (with very low doses), a few times a week, or just once a week.

As you pot up plants into larger containers, you'll be replacing germination mix with a more nutrient-rich potting soil. This is a good time to mix in granular fertilizer. Be aware that it can be very easy to overfertilize small pots with

Create Efficient Systems

SEEDLING TRANSPLANT ADAPTABILITY

EASY TO SEPARATE AS SEEDLINGS	DIFFICULT TO SEPARATE AS SEEDLINGS	SHOULD BE DIRECT SEEDED
Basil	Beans, edible soy (edamame)	Arugula
Broccoli	Beans, fava (broad)	Beets
Broccoli raab	Beans, lima	Carrots
Brussels sprouts	Beans, shell (fresh or dried)	Celeriac
Cabbage	Beans, snap	Cilantro
Cabbage, Chinese	Corn, sweet	Garlic
Cauliflower	Cucumbers	Lettuce, baby mix
Celery	Dill	Mustard greens
Chard, Swiss	Melon, cantaloupe, honeydew	Parsnips
Collards	Peas, shelling	Peanut
Eggplant	Peas, snap	Potatoes
Endive	Squash, gourds	Radishes
Fennel, bulbing	Squash, pumpkins	Rutabagas
Kale	Squash, summer	Turnips
Kohlrabi	Squash, winter	
Leeks	Watermelon	
Lettuce, head		
Mâche		
Okra		
Onions, bulb		
Pak choi		
Parsley		
Peppers, hot		
Radicchio		
Scallions		
Sweet potatoes		
Tomatillos		
Tomatoes		

a granular fertilizer, so be judicious with your applications. Overfertilization can lead to stress via salt buildup and can lead to fungal growth and damping-off. Consider applying a small amount of granular fertilizer (or none at all) and continuing a routine of liquid fertilizer applications for the plants' entire nursery experience.

Cuttings and grafted plants should not be fertilized until they are potted up out of the initial rooting phase. Plants are incredibly sensitive when undergoing these healing processes, and applications of any fertilizer is likely to cause much more harm than help. For more on fertilizers and their applications, see Chapter 9.

POTTING UP PLANTS

The need to pot up plants depends on the specific crop, your climate, and the variables of the particular season. Many crops can be germinated in shallow flats (plug trays, soil blocks, open flats) and then moved directly into the garden. Plants that may need to be potted up are typically long-season, heat-loving, and very cold-sensitive crops, such as tomatoes, peppers, and eggplant.

Standard nursery procedure is to germinate all crops in shallow flats, choose the healthiest individuals from each variety for potting up, and graduate these to 4-inch pots or larger for continued indoor grow out. Almost any crop should be ready for outdoor planting directly from a 4-inch pot unless you start your crops incredibly early in the season, at which point potting up into gallon-size containers might be necessary or desired to hold them inside for a longer period without getting rootbound and stressed.

HARDENING OFF TRANSPLANTS

Hardening off is the process of acclimating your transplants to the stresses of the outdoor environment. These stresses include cooler temperatures, direct sunlight, wind, and varying levels of soil moisture. Think of it as a fitness boot camp

for your plants. It ensures they experience less shock when being transplanted. Hardening off is most useful for starts that have spent their entire lives in a temperature-controlled greenhouse. If transplants have been grown in an unheated cold frame, they've already experienced low nighttime temperatures, so they need less acclimating.

To harden off your plants, move them outdoors from the greenhouse or propagation area early in the morning or in the late afternoon/evening. Avoid moving plants from a greenhouse into direct sunlight at midday. If freezing temperatures are not a risk, you can leave the starts outside in their containers for about a week before transplanting into the garden to let them experience full sun and cooler nighttime temperatures. If nighttime temperatures do drop below freezing, you should bring the starts indoors overnight. Cut back a bit on watering to let the plants experience drier conditions, but don't let plants wilt or become water stressed.

There are many different theories about how gradually plants should be hardened off, or if it's even necessary at all. Some growers keep their plants in a shady area for the first few days and then move them into full sun; some bring the transplants indoors on the first few nights of the hardening-off process. In our experience, working in a variety of climates, simply getting the starts outside for about a week is plenty to harden them off for transplanting. Additional coddling seems to be largely unnecessary.

MONITORING FOR PESTS & DISEASES

Greenhouses and indoor growing spaces can be prone to higher concentrations of pest and disease issues than outdoor spaces are. Keeping your space clean and well ventilated will go a long way toward reducing or eliminating problems. Whenever possible, open windows or vents for cross ventilation.

If fungal or disease issues arise, temporarily moving the plants outside (weather permitting)

Whitefly (*Aleyrodes proletella*) is a pest that's most often found in greenhouses.

and wiping down all surfaces with an appropriate concentration of alcohol, bleach, or hydrogen peroxide will help get things under control. Sticky traps and other low-tech pest catchers can help keep persistent insect populations under control. Allowing the soil surface in your flats to dry out between waterings will reduce the possibility of insect breeding.

INSECT PESTS

Whiteflies. These are a common greenhouse pest. If you brush over a plant and a cloud of very tiny white insects poofs up into the air, you have whiteflies. These insects slowly desiccate crops by sucking their sap. They're usually more annoying than harmful, but they can make plants more prone to disease problems. Sticky traps or soap spray is effective for killing them.

Aphids. These are tiny (⅛ inch long) pear-shaped bugs that hang out mostly on the underside of leaves and suck the sap out of your crops. They can be a problem in the nursery and in the garden. If you see leaves curling or looking anemic and yellow, check their undersides for aphids. Aphids are especially attracted to lush, green growth. Carefully rinse them off your plants with a hose-end sprayer, or spray them with a soap spray. Sometimes, moving nursery plants outside for several hours will dissipate infestations. If you have a serious infestation, it could be a sign that the plants are getting too much nitrogen, so cut back on your fertilizing.

Thrips. These are very difficult to see with the naked eye. Unless you have a magnifying glass, they just look like tiny dots moving around on your crops. Silvery spots or streaks (sometimes with black spots of fecal matter) on plants are caused by thrips scraping plant tissues and feeding on their juices. Soap spray or neem oil works well for controlling them in the nursery.

Fungus gnats. These are black bugs similar in size and appearance to mosquitoes. They're attracted to organic matter and moisture in your potting soil, and can often be seen hovering about the plants in a greenhouse. They don't cause much damage unless populations get out of control. They lay their eggs in the soil, so large numbers of larvae can damage plant roots. Keeping the greenhouse clean and well ventilated helps limit the growth of these pests. Allowing the surface of the soil to dry out between each watering can also significantly reduce their population. Yellow sticky traps and pyrethrum-based organic insecticides kill them, and beneficial nematodes will kill soil-dwelling larvae.

Proper cleaning and sterilization of nursery equipment will help ensure that damping-off never becomes an issue.

DISEASES

Damping-off. This is a fungal disease that attacks the stem of young seedlings. The top part of the plant will tip over, and the stem will appear to be "chopped" through completely at the base, or may be black and rotted. Proper ongoing cleaning and sterilization of nursery equipment will help make sure this disease never becomes an issue. Adequate ventilation also goes a long way toward preventing this disease. Using a small amount of high-quality compost in your potting mix will also help by maintaining a population of beneficial bacteria that compete with the damping-off fungus. If you see signs of damping-off, remove the affected plant or plants (burn diseased plants when possible or add to a hot compost pile), increase ventilation, and cut back on watering as much as possible. You can also try dusting cinnamon lightly on the soil around the rest of the plants, misting with a mild chamomile tea, or spraying neem oil—all of these have natural antifungal properties that can help combat the disease.

Botrytis. This is a fungal disease characterized by gray fuzzy mold. It thrives in damp, cool conditions, and is often a problem for tomatoes and strawberries grown in tunnels. It can also show up in minimally heated propagation areas. If you see signs of it, increase ventilation and consider using a fan to dry off foliage. If possible, raise the temperature in the propagation area above 75°F (24°C). Try spraying chamomile tea, a sulfur- or a copper-based organic fungicide, or an organic spray that contains *Bacillus subtilis* (such as Serenade).

Mustard greens can be grown as microgreens under an eight-bulb T5 fluorescent grow light.

year-round
nursery production

↓ Now that you've set up a nursery or propagation area, why not use it as a year-round greens factory? Growing sprouts and microgreens is a great way to supplement your garden production and to ensure a steady flow of fresh greens during winter when garden harvests are less prolific. For the serious home food producer with an adaptable propagation space (or even just a kitchen counter), microgreens and sprouts can be worth their weight in greens.

Any crop that has a totally edible structure can be grown as a sprout. Obviously, you'll have to avoid nightshades and any other crop with potential toxins, but feel free to experiment with different salad greens and herbs. Some unexpected crops like sunflowers and grains like wheat, buckwheat, and rye can produce delicious winter greens.

Sprouts are an easy way to grow salad greens indoors.

WHAT YOU'LL NEED

Glass containers. Sprouts can be grown in any large glass container. We use quart-size mason jars.

Sprouting lids. You can either purchase these or make them at home. The best lid will securely screw onto the top of your jar and provide a fine-mesh screen to allow water to easily drain from the container but hold back small seeds.

Holding rack. You'll need some sort of rack to support the upside-down jars and allow them to drain effectively. Jars can simply be placed upside down in a bowl or set in a dish-drying rack near the kitchen sink.

HOW TO SPROUT

The technique for growing most sprouts is very similar: Measure a few teaspoons (or 1 cup for large seeds) of seeds per quart jar, fill the jar with water, and let the seeds soak for 4 to 12 hours. The volume of seeds to use and how long they need to soak varies by crop. See the chart on page 202 for specific information on each crop.

After soaking, rinse the sprouts twice a day and leave the jar upturned to drain. Sprouts can begin germinating in as little as 12 hours after soaking, but most take 1 to 2 days. Sprouts grow quickly, and after 3 to 5 days, most varieties of sprouts will have filled the jar and should be placed in the refrigerator and eaten within 7 days.

When starting out, we recommend sprouting between one to three quart jars of sprouts per week for a household of one to three people. Make a schedule so you can continue with your successions of sprouts each week and maintain a steady supply. Because of the quick turnaround time, it is easy to ramp production up or down based on your actual use.

SPROUTING

There are dozens of commonly available seeds that are good candidates for sprouting, including alfalfa, mustard greens, radish, and fenugreek. A motivated gardener could collect enough seed from a few well-managed plantings of alfalfa and mustard greens to provide themselves with fresh sprouts through the winter months.

Sprouts can be grown any time of year with minimal effort and space. If you're purchasing seed for sprouting, make sure the seed has not been treated with any fungicides or other chemicals. Many seed companies sell seeds that are labeled as "sprouting seeds." Sprouting seeds should not be treated and should be certified pathogen-free, but this is not always the case. Labeling requirements are somewhat vague on this, so check with your supplier to be sure. Many co-ops and organic-focused grocery stores sell sprouting seeds in their bulk food section.

MICROGREENS

Growing microgreens is another great way to produce food year-round with a minimum of space and time. Microgreens are essentially just sprouts that are propagated in a nursery flat on a bit of soil and allowed to grow larger than traditional sprouts. These sprouts will be grown until they start to put on tiny leaves. As the name suggests, microgreens are just very tiny salad greens. You can grow flats of arugula, mesclun mix, mustard greens, and just about any other crop with edible greens. Wheatgrass can be grown as a microgreen for making a healthful juice.

WHAT YOU'LL NEED

Flats. Microgreens can be grown in virtually any tray or shallow container, but we prefer to use standard nursery flats with a few small, bottom drainage holes. These containers retain some water, which helps keep the soil moist; however, it's important to avoid saturating and drowning the greens.

Soil mix. You can use potting soil, germination mix, or even straight garden soil as a medium for the greens. Plan to compost or dump the soil into the garden after use, as it will be full of miniature roots that will need time and space to decompose. Fill each container with 1 to 2 inches of soil, about half the depth of a standard flat.

Lighting. Microgreens can be grown with natural light or under any grow light. Supplemental light helps the plants grow more quickly and evenly, but amazing microgreens can be grown on a windowsill. Four to 8 hours of supplemental light from your grow light system will make the greens grow faster and straighter.

GROWING MICROGREENS IS A GREAT WAY TO PRODUCE FOOD YEAR-ROUND WITH A MINIMUM OF SPACE & TIME.

HOW TO GROW MICROGREENS

The quantity of seed needed for each flat of microgreens depends largely on the size of the seed itself. Generally, larger volumes of seed are needed for larger seeds, although the total number of seeds may be less. For example, a flat of the relatively large-seeded cilantro may use between 1 and 2 tablespoons per flat; whereas a crop with a smaller seed like mustard greens may use 1 teaspoon per flat. Use the chart on page 202 and your own personal trials to figure out the volume of seed that works best for you.

Microgreens are wonderfully easy to care for. They don't require any fertilization, although we do prefer to use a soil mix with compost to ensure the plants have access to some nutrients as they grow. We typically use overhead watering for flats of microgreens until they have germinated and then switch to bottom watering once they have emerged. Bottom watering prevents the thin, tender sprouts from being knocked over and getting dirty.

After 1 to 3 weeks, depending on crop, temperature, and light levels, your greens should be ready to harvest. It's standard practice to cut them when they just begin to show their first true leaves, but you can cut them at any size you like.

When starting out, we recommend sprouting one or two flats of microgreens per week for a household of one to three people. Make a schedule so you can seed flats once per week and maintain a steady supply. Like sprouts, it's easy to ramp production up or down based on your actual use.

CROPS FOR SPROUTING & MICROGREEN PRODUCTION

Many crops can be grown as sprouts and microgreens, but some are better suited to growing this way than others. For example, any brassica can be easily sprouted. However, broccoli and cauliflower seed are both more expensive and slower growing than mustard greens, so you may prefer to use mustard greens for sprouting and microgreens.

This chart contains our suggestions for the easiest and best crops for sprouts and microgreens. Some of these crops may be unfamiliar to you (like fenugreek) and others may be familiar (sesame, millet). Even though these are not typical vegetable crops for the home gardener, they are readily available at grocery stores and work well when creating delicious sprouting and microgreen mixes.

CROP	SPROUTING	QUANTITY OF SEED*	HOURS TO SOAK	DAYS UNTIL HARVEST	MICROGREENS	QUANTITY OF SEED**	DAYS UNTIL HARVEST
Alfalfa	Y	3 tablespoons	6	5–6	Y	1.5 teaspoons	10–14
Almonds	Y	1 cup	1	1			
Arugula	Y	3 tablespoons	6	4–5	Y	1.5 teaspoons	10–14
Basil					Y	1.5 teaspoons	10–14
Beans, edible soy (edamame)	Y	1 cup	12	2–4			
Beans, fava (broad)	Y	1 cup	12	2–4			
Beans, lima	Y	1 cup	12	2–4			
Beans, shell	Y	1 cup	12	2–4			
Beans, snap	Y	1 cup	12	2–4			
Beets					Y	1.5 tablespoons	10–14
Broccoli	Y	3 tablespoons	6	4–5	Y	1 teaspoon	10–14
Buckwheat	Y	1 cup	6	3–4			
Cabbage	Y	3 tablespoons	6	4–5	Y	1 teaspoon	10–14
Carrots					Y	2 teaspoons	10–14
Celery					Y	1.5 teaspoons	10–14
Chard, Swiss					Y	1.5 teaspoons	10–14
Chickpeas	Y	1 tablespoon	12	2–3			

*Per quart jar **Per standard flat

CROP	SPROUTING	QUANTITY OF SEED*	HOURS TO SOAK	DAYS UNTIL HARVEST	MICROGREENS	QUANTITY OF SEED**	DAYS UNTIL HARVEST
Cilantro					Y	1.5 tablespoons	10–14
Clover	Y	3 tablespoons	6	4–5			
Corn, sweet	Y	1 cup	3	2–3	Y	2 tablespoons	10–14
Cress	Y	3 tablespoons	4	3–5	Y	1 teaspoon	10–14
Dill					Y	1.5 teaspoons	10–14
Fenugreek	Y	½ cup	8	3–5			
Kale	Y	3 tablespoons	6	4–5	Y	1 teaspoon	10–14
Leeks					Y	1 teaspoon	10–14
Lentils	Y	1 cup	12	3–4			
Lettuce, baby mix					Y	2 teaspoons	10–14
Millet	Y	1 cup	8	2–3			
Mung beans	Y	1 cup	12	3–5			
Mustard greens	Y	3 tablespoons	6	3–5	Y	1 teaspoon	10–14
Oats	Y	1 cup	12	2–3			
Parsley					Y	2 teaspoons	10–14
Peas, snap	Y	1 cup	12	2–3	Y	¼ cup	10–14
Peppercress	Y	3 tablespoons	4	3–4	Y	1.5 teaspoons	10–14
Purslane					Y	2 teaspoons	10–14
Radishes	Y	1 cup	6	4–5	Y	1 tablespoons	10–14
Rye	Y	1 cup	12	2–3			
Scallions					Y	1 teaspoon	10–14
Sesame	Y	1 cup	6	1–2			
Sorrel					Y	1.5 teaspoons	10–14
Spreen					Y	3 teaspoons	10–14
Squash, pumpkins	Y	1 cup	8	1			
Sunflower	Y	1 cup	8	1–2	Y	2 tablespoons	10–14
Turnips	Y	3 tablespoons	6	4–5	Y	1 teaspoon	10–14
Watercress	Y	3 tablespoons	6	3–5			
Wheat	Y	1 cup	12	2–3			

CHAPTER THIRTEEN

IRRIGATING CONSISTENTLY & EVENLY

Water is one of the most important factors that determine the success or failure of a crop. Like most living organisms, plants are mostly water, and many fruit and vegetable crops contain an even higher percentage of water than other kinds of plants. Providing your crops with adequate and consistent moisture can dramatically increase yields and improve crop quality.

Your garden's watering requirements will depend on a variety of factors, including the kinds of crops you choose, the structure of your soil, your region's climate, and the day-to-day variations in your weather. Rainwater is a great source of hydration for your plants, but it doesn't always arrive at the required time and in the required amount to sustain needy vegetables. In general, crops rarely yield reliable and robust harvests without supplemental watering at some point during their life.

Drip irrigation efficiently delivers water directly to the roots of your crop.

the case for irrigation infrastructure

We strongly encourage vegetable gardeners to include a watering system with a timer as a part of their initial garden plan. Keep in mind that, just like any other element of your project, even a perfectly built automatic irrigation system will still need ongoing attention. Making adjustments to the watering duration and frequency throughout the season will save water and ensure that your crops get exactly the amount of moisture they need. Additionally, normal wear and tear will inevitably cause small breaks and leaks that you'll need to repair. Even with these tasks, though, your total time spent dealing with water issues will be reduced by 90 percent in comparison to hand watering.

An automatically timed watering system will save you countless hours managing a hose or otherwise attempting to keep soil moist by hand. It may even allow you to take a vacation once in a while. Because a timed system can be

THE RIGHT AMOUNT OF WATER

Most vegetable crops need at least 1 inch of water per week, whether from rain or irrigation. This can rise to 2 inches in arid climates or during hot weather. The best schedule to deliver that amount of water each week will depend on your climate, soil type, and crop selection. Standard procedure is to water crops deeply every few days. The idea is to promote the growth of large root systems, encouraging the plants to chase the water deep into the subsoil. However, in sandy soil, you will likely need to water more frequently and for shorter durations. The most important thing is that the plants appear healthy, continue to grow at an appropriate rate, and produce abundant harvests.

Gardeners often compare the soil to a sponge. When it dries out completely, it turns rock hard, and it takes some effort to convince it to absorb water again. If it stays completely saturated with water, it becomes a funky mess. The goal is to keep soil moisture in the "golden range" for plant growth. This is comparable to the feel of a damp sponge after you've wrung it out. The standard test is to squeeze a handful of soil in your palm: If it holds together in a ball, it has the right amount of moisture. If it falls apart, it is too dry. If you drop the ball of soil from waist height, it should shatter; if it doesn't, the soil is too wet.

The goal is to irrigate the soil until it is saturated, let it dry out enough to reduce habitat for fungus and mold, and then water it again. You never want to see your crops wilt due to lack of water. A wilting crop is water stressed, and this can lead to low yields and susceptibility to disease. We highly recommend that you develop a feel for your own soil, to learn its tendencies: what it looks like when it's too dry, too wet, and just right. Feel your soil when it's really dry, after watering, and after a heavy rainstorm. Taking the time to understand your soil will allow you to easily and quickly identify its watering needs and keep plants on track.

set to deliver water regularly and at the best time of day for irrigation, your plants will be healthier and will produce a larger yield. Building the system into the garden at the outset will save valuable hours when compared to creating an "aftermarket" solution that must be delicately placed around all of your established structures and plants.

Because we strongly believe that hand watering a large, diversified garden site is an inefficient use of time and resources, we won't even include it as a viable option for garden irrigation. In the peak of the season when irrigation is most commonly needed, spending valuable hours trailing a hose through the garden is, at best, a poor use of time. Not to mention the likelihood of accidentally crushed plants. We've destroyed our fair share of bush beans trying to pull an unwieldy hose around the corner of a bed, and we have learned from the experience.

With an established irrigation system in place, these hours can be much better used identifying and managing pest and disease problems, harvesting crops, and succession planting new crops in the garden.

The simple act of automating your watering program will completely change the health, yields, and appearance of your garden. However, this doesn't mean that it's time to throw out all of your hoses and spray nozzles; those are still essential tools for spot watering the garden and watering in new crops as you seed or plant them in the garden.

Each bed in the garden receives drip irrigation from mainline tubing run underground and drip tapes placed across the surface of the soil.

components of an irrigation system

↓ We like to visualize our irrigation systems from the top down, so let's look at the various parts you'll need, starting at the water source and working our way into the garden beds. As you identify where these parts will go in your system, you can label them on your irrigation map (see page 217) for future reference. This will help tremendously when you're placing a parts order or heading to the store. Because running an irrigation system directly off an outdoor hose spigot is typically the most straightforward option for home gardeners, we'll primarily focus on components for that type of system.

FITTINGS FOR CONSTANT PRESSURE

Some irrigation components are designed to be used under constant pressure (with water pressure against them at all times), and some are not. In most cases, the backflow preventer, the Y-valve or manifold, and the timer are the only components that need to be rated for constant pressure. All other components should be placed *after* the timer, so they're only under pressure when the system is running. Failure to do this can result in a nasty leak, which always seems to happen when you're out of town.

Backflow preventer/antisiphon valve. This piece ensures that water will not come backward into your home water pipes from the irrigation system. It empties out when the water spigot is not in use, to prevent reverse siphoning. Make sure the piece has hose threads, not pipe threads, on both the inlet and outlet sides, so you can screw it right onto your spigot. Hose threads are labeled FHT or MHT (female hose thread or male hose thread) and pipe threads are labeled FPT or MPT (females pipe thread or male pipe thread).

Many newer water spigots already have a built-in backflow preventer, in which case you don't need to add one. A built-in backflow preventer might look like a small cylinder with holes in the side attached to the outlet of the spigot, or a plastic or metal "cap" on the top of the spigot. If in doubt, note the model name and number of the spigot and look online or give your local plumber or plumbing store a call.

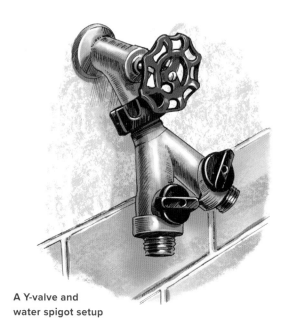

A Y-valve and water spigot setup

Y-valve or manifold. Using a Y-valve or other line splitter will give you the option to leave the irrigation system connected all season while still using the hose. This prevents you from forgetting to reattach the timer after you finish using the hose. Keep in mind that many hose thread Y-valves and manifolds are limited in the volume of water that can pass through them. If you are running a lot of high-volume sprinklers or an enormous drip system, look for a "full flow" valve.

Timer. For most home vegetable gardens, a battery-operated timer with hose threads is the best option. They are cost effective, easy to set up, and don't require that the water source be located near a power outlet. Larger, plug-in timers are great if you're creating a system with multiple irrigation zones (see page 210). Installing such a timer and associated system is a more complicated procedure, and is beyond the scope of this book (see Resources, page 292, for books on complex irrigation system design). If you want to set up multiple zones in your irrigation system, there are hose thread battery timers available with multiple outlets. You can also use a manifold on a water spigot and multiple timers to create as many zones as you like. Many hose thread timers are limited in the volume of water that pass through them, so if you are running a lot of sprinklers you may need to upgrade to a zoned system with a remote timer.

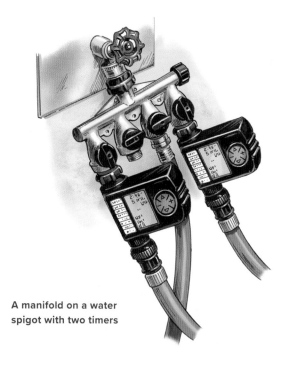

A manifold on a water spigot with two timers

PRESSURE REGULATOR/ PRESSURE REDUCER

This fitting reduces the water pressure from the water spigot to a level that is appropriate for your drip lines. If the water pressure in the system is too high, drip lines and fittings can blow out or shoot water up in the air rather than drip it slowly into the soil. Be sure to choose a regulator with a psi (pounds per square inch) rating that matches the psi required by your drip lines. If you're irrigating with sprinklers, it's unlikely that you'll need to reduce your water pressure.

FILTER

Using a filter will catch tiny particles that may be in your water and help prevent drip lines and emitters from clogging. Filters are usually rated in "mesh" sizes; the higher the number, the finer the screen. Most drip lines need somewhere between 100- and 200-mesh filtration. Make sure to choose a filter that's rated for your drip lines. Remember to clean out the screen on the filter at least once or twice a season. Unless you're irrigating from a well with lots of particulates, you won't need a filter when irrigating with sprinklers.

MAINLINE/HEADER TUBING/SUPPLY LINE

The mainline is a solid tube that brings water from the spigot to the garden. For very simple irrigation setups, you can use a garden hose for your mainline, but we suggest polyethylene (often referred to as "poly tubing") for all applications. In an irrigation system, mainline tubing is superior to a regular hose because it is inexpensive, lightweight, and easy to cut to fit your spaces. Most systems use ½- or ¾-inch polyethylene tubing for the mainline. Polyethylene is also a relatively stable plastic, UV-resistant, and unlikely to leach chemicals into your water and soil.

It's important that the brand of your fittings match that of the tubing. Even if two brands are labeled the same diameter, the true size of the pieces can vary by fractions of an inch, making them difficult or impossible to integrate. You'll need an assortment of fittings for your system, the type and number of which will be indicated on your irrigation map (see page 217). Here are a few essential pieces.

+ **Header start fitting.** This connects the mainline to the threaded end of the filter or pressure regulator.

+ **Header end fitting.** This fitting closes up the ends of each branch of your mainline tubing. These may be figure eight fittings, flush caps, or another variation on an end piece.

+ **Tees, elbows, and crosses.** You'll want to pick up as many of these fittings as necessary to turn your mainline around corners, and to branch it out through the garden. Order several extra of each, just in case.

+ **Couplers.** Make sure to pick up a handful of straight couplers to attach the ends of two sections of mainline tubing together and to patch areas that you accidentally puncture or cut too short.

+ **Valves.** These are great for isolating different parts of your irrigation system. Many growers will put a valve on each bed, so that if one is empty or the crops in it don't need extra water, they can shut it off. Some growers put a valve on each drip line for even more precise control. You can add a valve to each sprinkler in an overhead irrigation system and at the end of the system to make it easier to drain.

STAKES

Stakes are incredibly helpful when trying to hold mainline tubing in place en route to the garden and inside individual beds. Look for irrigation stakes or any type of landscaping pins that will fit your system. Drip lines are often light enough that they can be blown around by the wind if not secured in place.

drip irrigation

↓ Applying water directly to the soil surface is the most resource-efficient way to irrigate your plants. Drip irrigation allows you to water the portions of the garden planted in crops, reducing weed growth in bed edges and pathways. Drip lines also make planting in straight rows easier since you can plant directly along the side of the line. They can also make it easier to space your plants evenly. For example, a drip system with emitters spaced every 12 inches will make a perfect guide for planting kale or any other crop you might want to space at 12 inches—simply put a transplant right next to each emitter. A well-functioning automated drip system, meaning there are no leaks, can deliver the appropriate amount of water with very little waste.

In our experience, one of the biggest drawbacks to drip irrigation is the increased difficulty of, and extra time spent, weeding if you are using a hoe (it's much slower to work around the plants without accidentally damaging the drip lines). In addition, it's more difficult to germinate closely spaced direct-seeded crops like lettuce mix and carrots.

Although drip irrigation systems typically take longer to set up than overhead systems, they're much easier to install and manage than you might think. All you need is a free weekend, a working water spigot (a.k.a. hose bib), a few tools, and a handful of supplies.

IRRIGATION ZONES

If your garden is very large, you might not have enough flow coming out of your water spigot to water the entire space at once. Splitting the garden into different zones—areas that are watered on individual schedules—will solve this problem. Having different zones can be useful for other reasons, as well. One area of your garden might get more sun and will need more water. Or you might want to put annual and perennial crops on different watering schedules.

If you're a whiz at irrigation and you have an existing underground irrigation system for your property, you may be able to adapt one or more of these zones into an effective drip irrigation system for your garden. Using this existing resource will eliminate the need to run a system off the water spigot and can save time and energy. If you can find a zone that can be completely redirected to the garden, you're in luck.

Keep in mind that drip irrigation and pop-up lawn sprinklers are not compatible. The pop-ups take too much water pressure for the drip to work properly, and drip irrigation and sprinklers usually require very different run times. Try to isolate the garden from the rest of the system or change the rest of the zone to drip lines as well.

DRIP LINES & EMITTERS

Drip lines come in a few different forms, the most common being tubing and tape. They can come either with water emitters already built in at regular intervals or with bubblers, which are meant to be manually inserted into the system wherever you want them. The advantage of using built-in emitters is the simplicity of setup and evenness of watering throughout the system. The advantage of using bubblers is the ability to highly customize the system to place water only where you want it. A system can be built using both of these materials but, generally speaking, we've found that drip lines with built-in emitters are the best choice for annual vegetable gardens.

Each type of drip line or emitter system has its own complement of fittings. These include fittings to connect it to the mainline, end fittings, tees, elbows, crosses, couplers, and valves. As you fill in your irrigation map, the locations for these pieces should become clear.

All drip lines have a limited length they can be run from the mainline. Be sure to follow the guidelines for your chosen drip lines to prevent dry spots at the far reaches of the system.

EMITTER SPACING

Drip lines are available with many different emitter spacings (4, 6, 8, 12, and 24 inches are common). Four- or 6-inch spacing works well for germinating direct-seeded crops, but uses more water than necessary for established and widely spaced crops. Spacing emitters 24 inches apart is good for perennials, but is not ideal for annual vegetable gardens. We find 8-inch spacing works well for most applications. Twelve-inch spacing is also effective, as long as you give closely spaced and direct-seeded crops a little extra care with the hose.

PRESSURE COMPENSATION

The emitters in many drip lines and bubblers are rated as "pressure compensating." Since pressure in a system will be lower farther from the water source, these emitters compensate for that difference and ensure all emitters drip at the same rate. Thus, a drip line located farther away from the water source will get the same amount of water as the lines closest to the source. It also allows for longer runs of drip line. If you're using a drip system without pressure compensation, you run the risk of unbalanced water distribution. The beds closest to the water source will get the lion's share of the irrigation. This is a particularly likely outcome in larger gardens.

TYPES OF DRIP IRRIGATION

Drip tape. These are flat, straight drip lines that are very cost effective and quick to set up. They are not flexible, so they'll only work for setting up straight rows. Drip tapes are relatively easy

DRIP OR OVERHEAD IRRIGATION?

In general, there are two types of systems to choose from: drip irrigation and overhead irrigation. Some growers employ both systems in their garden. We've found that for most small to midsize home gardens that include a diverse range of crops, drip irrigation does the best overall job with the least amount of resources. Overhead irrigation excels at germinating closely spaced direct-seeded crops and is very efficient to manage if all of your crops are short or of medium height. However, in a home garden with crops of many different heights, we find that tall crops often block sprinkler coverage for shorter crops. In addition, sprawling crops with large leaves like squash and melons need a lot more irrigation time with a sprinkler to get water to the roots than compact crops like lettuce or carrots.

to damage while gardening, but also very easy to repair if leaks do occur. They may come in variations with different flow rates and different wall thickness. Lower flow rates will allow you to set up a larger system without the need to break it into zones, but this is not usually an issue except in very large gardens. Higher flow rates will deliver the desired amount of water in a shorter period of time. We typically use thicker drip tapes with higher flow rates. Our preferred tapes are 15 mil thick and rated as "high flow."

Quarter-inch emitter tubing. These small hoses are very useful for container irrigation and beds with nonlinear shapes (circles, ovals, etc.). They are relatively easy to set up, although the small fittings can be challenging to put together. They are more durable than drip tapes, but not typically pressure compensating. They still work great as long as you don't exceed the specified run length from the mainline.

Clockwise from top left: Quarter-inch emitter tubing irrigates mustard greens. Cilantro is seeded along a drip tape. Bubblers can be used to deliver water directly to each plant. Half-inch emitter tubing can be run in straight or curved lines.

Half-inch emitter tubing. These are very burly and long-lasting drip lines. The heavier tubing is more expensive than other options and can be somewhat awkward to work with, but once in place will last virtually forever. We really like these for perennial crops that will be in place for a few years. Some growers use them for all their irrigation needs, including on annual crops.

Bubblers. It's possible to set up solid tubing and insert your own emitters at the spacing and locations that you prefer. This type of system requires more setup time and can be more expensive, but some gardeners find the customization worth it. It can be especially useful when setting up a perennial bed with crops at different spacings that will be in place for a few years. Pressure-compensating bubblers will be significantly more expensive, but worth it for large spaces.

Black porous soaker hose. If at all possible, avoid using black porous soaker hoses. They are notorious for watering unevenly, they degrade in quality quickly, and they are easy to break when working in the garden.

MANAGING THE DRIP SYSTEM

It's a good idea to manually run the irrigation system every week or two while you're working in the garden. This allows you to check for new leaks and make sure the system is functioning properly. It's also a good time to assess if the system should be turned up or down to accommodate weather patterns and soil moisture levels.

The need for irrigation will vary drastically over the course of the season. It may be that watering 1 day a week is sufficient at certain times of year and that 7 days a week may be necessary at other times. Once the system is in place, management will take at most a few minutes a week. Simply adjust the settings on the timer and carry on with your other essential garden tasks.

Keep in mind that with drip irrigation, your plants and seeds may benefit from a little extra help when first planted and germinating. The roots of seeds and newly transplanted crops are near the soil surface, and can't easily reach the subsurface moisture the drip system provides unless they're immediately adjacent to an emitter. This is the most sensitive part of their life, and it's essential that they don't dry out.

We suggest always watering in new transplants and seeds with a hose or sprinkler. Depending on the weather and time of year, you may want to hand water a few more times, or at least check soil moisture more frequently during the first week or two after planting to make sure crops get off to a healthy start. Some growers will decrease their watering duration and then run their drip system every day for 1 to 2 weeks if germinating seeds during hot, dry periods. This prevents the seeds from drying out. These little bits of extra effort can pay huge dividends in improving germination rates and early plant growth and development.

DURATION OF IRRIGATION

The length of an irrigation session will vary widely depending on the structure of your soil, flow rate (GPH, or gallons per hour) of your drip lines, and weather. Here is a starting guide for setting the system to run at the beginning of the irrigation season.

+ Drip tapes, low flow (approx. 20 GPH per 100 ft. of tape; 8-in. emitter spacing): 1 hour, four times per week.

+ Drip tapes, high flow (approx. 40 GPH per 100 ft. of tape; 8-in. emitter spacing): 30 minutes, four times per week.

+ Quarter-inch emitter tubing (½ GPH per emitter): 30 minutes, four times per week.

+ Half-inch emitter tubing (with ½ GPH emitters): 30 minutes, four times per week.

+ Bubblers (each emitter will have a designated GPH rating): 30 minutes, four times per week.

After your first watering session, finger test the soil. If it seems dry, add 10 to 20 minutes to your watering cycle and test again after your next session. Add length to the sessions and add extra watering sessions as the season progresses, as necessary.

Soil that appears dry may be moist just under the surface, and vice versa. Probing the soil with your finger will give you more information.

overhead irrigation

Some growers prefer the evenness of watering with an overhead sprinkler. They feel that it better mimics the natural effect of rainwater, which plants evolved alongside. Overhead watering is especially beneficial for newly seeded crops while they germinate. The complete coverage and delivery to the top layer of soil can help ensure a good germination rate, which saves time later on reseeding thin crop stands. Overhead watering also has the advantage of being simpler to set up when compared to a drip irrigation system. In addition, overhead watering does not interfere with weeding around your crops. You won't have to move drip lines out of the way or worry about accidentally cutting them with your weeding implements. Many professional market farmers use overhead irrigation for this reason.

Drawbacks are that overhead watering systems tend to use considerably more water than a drip system and they tend to water areas (including pathways) that don't actually require irrigation. This can lead to additional weed growth which, in turn, leads to extra time spent weeding. With certain crops, watering the entire plant rather than just the soil and root zone can encourage disease problems. Because it is much easier to overwater with a sprinkler system, their use may also lead to faster leaching of nutrients from your soil, and the large, heavy droplets can compact bare soil.

When using overhead irrigation, plan to water in the morning, as early as possible. This allows the water to soak thoroughly into the soil and gives plants a chance to dry off before the sun is at its full effect. Watering in the middle of the day is less efficient, because water is lost to evaporation. Watering in the evening can increase the prevalence of fungal disease, since the water sits on the leaves of your crops and the surface of the soil throughout the night.

CHOOSING THE RIGHT SPRINKLER

The most important consideration when purchasing and setting up a sprinkler is whether it is capable of adequately covering your garden space and whether it will be able to evenly cover the space.

Oscillating. These are sprinklers that move a fan-shaped spray of water back and forth over an area. They're very inexpensive and easy to set up, and can be moved quickly around the garden if you need to water more than one space in a single watering session. However, they usually put out much more water than the soil can absorb, so are prone to wasting water and can cause erosion and soil compaction if not carefully used.

Impact. Many small farmers who use overhead irrigation will choose impact sprinklers—those that shoot a narrow jet of water in a circle around the unit. These are more efficient at using water than oscillating sprinklers. When placed on a tall post, they can distribute water over your crops in a 360-degree radius. They also have the advantage of generally being easier to adjust to fit your needs, as they can be set to water in a narrow range all the way up to a complete circle. A single impact sprinkler can cover a lot of area but needs relatively high water pressure and volume to function properly.

Wobbler. These sprinkler heads can operate with low water pressure. They distribute water gently—closely mimicking a soft rain shower—and provide even distribution. They are typically designed to water in a 360-degree radius, so are best placed in between beds rather than on the edge of a garden site (as may be done with adjustable range impact sprinklers).

Microsprinklers. Microsprinklers function well with low water pressure and volume. They are a great option for setting up an overhead irrigation system in a propagation greenhouse, where they can be hung over your seeding flats from the greenhouse structure. They can also be effectively used for outdoor beds, but keep in mind they spray horizontally so need to be placed on risers to spray over tall crops. You can choose sprinkler heads with different spray patterns (360 degrees, 180 degrees, 90 degrees, etc.) to work with the shape of your beds.

SPRINKLER SETUP

With any overhead irrigation, the key is to space the sprinklers properly so that all areas receive adequate coverage. Systems can be built with permanent or semipermanent piping or can simply be attached to a garden hose. If you have a larger site, consider laying out pipes that can stay in place all season with risers at regular intervals. This will eliminate the need to move sprinklers and drag hoses between beds, potentially damaging crops. Depending on your climate and storage capacity, pipes can be drained and left in place over winter or brought inside for protection. The spacing of risers or stands will depend on the exact sprinkler head you buy and your water pressure. Good irrigation suppliers will help you plan for the correct fittings, mainline tubing, spacing, and number of sprinkler heads you'll need.

HOSE-END SPRINKLERS

If you want to start simple and use garden hoses, simply attach a hose thread timer to the water spigot that is nearest to your garden, attach a hose to the timer, and then attach the sprinkler to the business end of the hose. Before starting, make sure you have enough hose to reach from the spigot to the farthest reaches of the garden space.

Find the right location for the sprinkler in the garden. If your garden is an odd shape, you may

Wobbler sprinklers are easy to set up and mimic the effect of rainfall.

have to move the sprinkler to a few different locations to determine where you'll receive the best coverage. Depending on the strength of the sprinkler, you may have to water the garden in two or three sessions, moving the sprinkler each time to achieve even coverage.

Alternatively, you can set up a few separate sprinklers and set them to slightly different times so they don't compete for water pressure. Setting up a system with more than one sprinkler is not hard, but does take a little extra setup time and materials. To run multiple sprinklers in the garden, set them to run in succession, so that the second starts shortly after the first has finished and the third starts shortly after the second has finished, and so on. In order to do this, you'll need either a timer with multiple outputs, so that several hoses can be attached to the source at once, or the system will need to branch out in the garden with timers attached at each junction.

Some sprinklers have inlet and outlet hose threads and can be set up to run in series; this kind of setup is only feasible if your system has enough water pressure. Typical water pressure in a home water line is between 40 and 70 psi. The pressure can vary greatly depending on whether

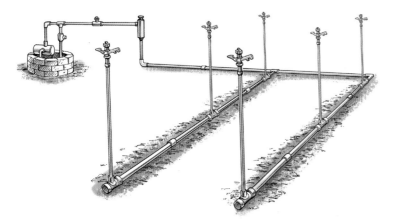
Impact sprinklers set up on posts are a good way to water tall crops.

or not other appliances (dishwasher, shower, etc.) are in use and even depending on water use throughout the neighborhood. The simplest way to determine if you are able to run a system is to set the sprinklers up and test them.

PERMANENTLY PIPED SYSTEMS

Most permanent and semipermanent systems are built using a mainline of either flexible polyethylene tubing or rigid PVC components. Depending on the size of your garden, water pressure, and the number of sprinklers you want to use, the mainline is typically ¾- or ½-inch tubing. Poly tubing has the advantage of being easier to disassemble, repair, and move. Risers can either be DIY-built from PVC tubing or purchased. PVC stands are much cheaper but may require additional work to stabilize and set in place.

Sprinkler systems like this work best in very geometric gardens, where it's possible to evenly space risers and achieve uniform coverage without ever moving a piece of the system. For example, a 60- by 60-foot square garden could be irrigated with two parallel lines of sprinklers set 20 feet from each outer edge with risers every 20 feet—a total of four sprinklers. In this scenario, you'd want to calculate your water pressure and make sure to purchase the correct sprinkler heads to deliver water 20 feet in each direction.

Also take into account the height of the riser and the angle of the sprinkler head itself. See Resources (page 292) for information on overhead irrigation suppliers.

WATERING DURATION & FREQUENCY FOR SPRINKLERS

Appropriate run times for sprinkler systems will vary greatly depending on the type of sprinkler you choose, so consult with your supplier for a starting point on how long to run your system. In general, oscillating sprinklers should be run for a much shorter duration than impact or wobbler sprinklers. In the absence of professional guidance, run a watering session to determine an appropriate duration for your sprinkler to run. Try starting with a 15-minute irrigation session, then finger test the soil and determine if the time should be extended. To perform the finger test, poke your finger into the soil—if it feels moist about an inch down, you've watered sufficiently. If it's wet on the surface but dry below, you need to water for a longer duration. If water is pooling on top of the soil, wait until it has absorbed before watering again. You want the soil to feel moist but not overly saturated.

Remember that you will necessarily be adjusting the duration and frequency of irrigation sessions throughout the season. With modern soil moisture meters and smart controllers, it's possible to create a mostly self-operating irrigation system. These technologies can save water and time; however, given the costs, they only make sense for growers who operate on a very large scale or know that they don't have the time to be regularly present in the garden. Even with advanced systems, it's a good idea to keep a watchful eye on soil moisture, as no system is completely reliable. Most growers still manage their system manually and check the soil moisture every few days to decide if the system should be adjusted. Remember to turn your system off during spells of rain and to take it apart before freezing weather sets in.

making an irrigation map

↓ We highly recommend drawing out your irrigation system. Photocopy or trace a copy of your garden map (see page 47) and use it to sketch out the path of the mainline. Drawing the whole system out on paper gives you the opportunity to think through design choices and make sure you are taking the easiest and least obtrusive path from the water source to the garden. For an overhead system, it'll allow you to determine the best locations for sprinklers in advance, without the trial and error of moving equipment around to see what gets coverage.

A SINGLE-TIMER SYSTEM

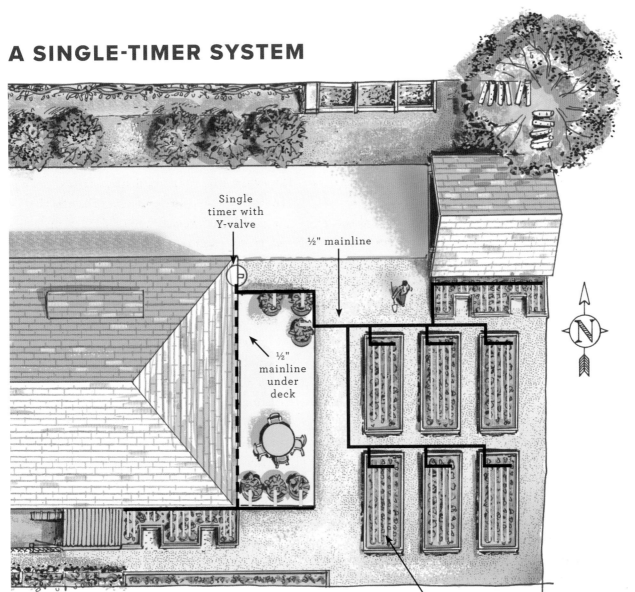

Single timer with Y-valve

½" mainline

½" mainline under deck

N

¼" emitter tubing

SEPARATE SYSTEMS FOR FRONT YARD & BACKYARD

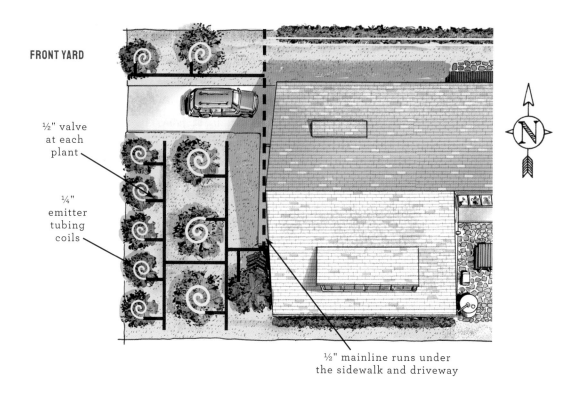

FRONT YARD

½" valve
at each
plant

¼"
emitter
tubing
coils

½" mainline runs under
the sidewalk and driveway

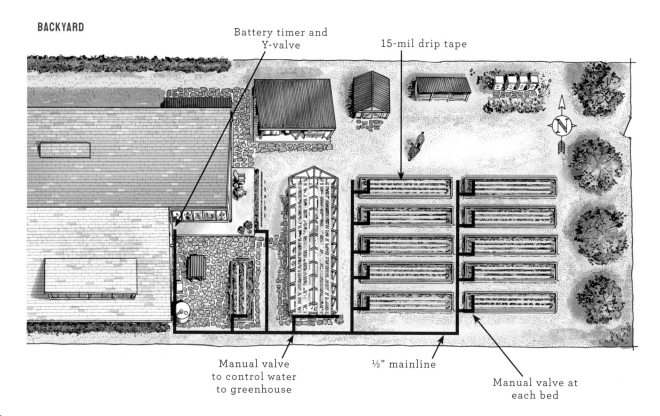

BACKYARD

Battery timer and
Y-valve

15-mil drip tape

Manual valve
to control water
to greenhouse

½" mainline

Manual valve at
each bed

START WITH THE MAINLINE

Start by identifying the nearest usable water source to the garden. You may be lucky enough to have a spigot in the garden area itself or very close nearby, or you may need to connect to a spigot on the house and run the irrigation to the garden. Look for the path of least resistance.

The mainline will need to be run in an uninterrupted line from the water spigot all the way to your beds. We find that it is often possible to run the tubing right along the foundation of the house until an opportunity arises to turn the tubing toward the garden. We like to follow the edges of existing elements like patios and decks to keep the tubing out of the way and make it easy to locate in case something goes wrong in the future.

In regions with mild winters, shallowly burying the mainline is the best way to eliminate the possibility that someone will trip over, mow over, or otherwise be annoyed by the tubing. Light freezes won't affect the tubing in winter because most of the water drains out through the drip lines when the system isn't running. If you live in a region with very harsh winters, consider burying the mainline below frost depth or bringing it indoors in winter so intense cold doesn't compromise the longevity of the plastic. Here in the Pacific Northwest (USDA Zone 8), we usually bury polyethylene mainline about 4 inches deep and leave it in place year after year.

A TWO-TIMER SPRINKLER SYSTEM

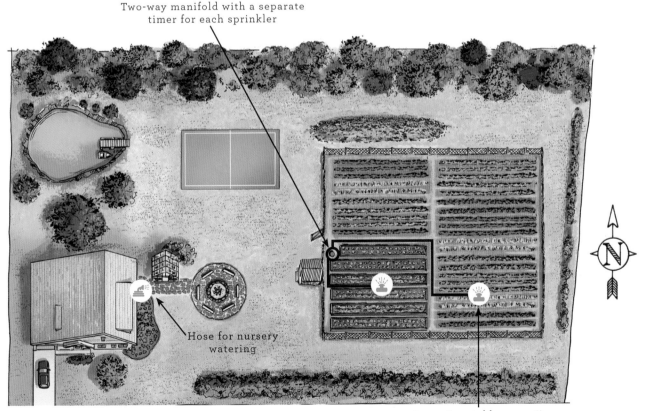

Two-way manifold with a separate timer for each sprinkler

Hose for nursery watering

Impact sprinkler on a 4' stand, covers one quadrant

Punch holes into mainline tubing.

Add tape locs to mainline tubing for drip tapes.

Connect drip tape to the tape loc.

MAP THE BRANCHES

Once you've identified a path to the garden, map out how the system will branch out to reach all of the beds. You can use tees, elbows, and crosses as necessary to get the mainline where it needs to go. Draw the path of the tubing as it reaches each bed; take the shortest path with the least amount of branching to make your job easier. Consider including valves on each bed, sprinkler line, or drip line so you can shut off the water to them if necessary.

If using drip lines, draw their arrangement within the beds. Drip lines should be set up to provide even coverage of the entire square footage of each bed. Since you'll be planting different crops at different spacing over the course of the year and from year to year, you want the drip lines to provide water for all crop spacings. This way you don't have to retrofit the system midseason or from year to year. If you already have crops in place, avoid the temptation to run lines directly to each plant; you'll only have to redo everything next year when you plant something different.

We usually space drip lines approximately 12 inches apart within a bed. Remember that the water dripping from each emitter will spread out in an 8- to 12-inch circle below the surface of the soil (larger circles in clayey soil, smaller circles in sandy soil). Emitter spacing on most drip lines is between 8 and 12 inches, so spacing the lines 12 inches apart will provide water to the entire bed. In addition, consider which direction is easiest to orient the lines and how this will influence your planting arrangements and access to the beds. Generally, running the mainline parallel to the shorter length of the bed and running the drip lines parallel to the longer length is the best way to go. For example, in a 4- by 8-foot bed, we run a drip line 6 inches from the bed's long edge, another one 12 inches toward the center, a third one 12 inches from that, and a fourth line 6 inches from the opposite edge.

Use snips to cut drip tape to length.

Fold the ends of drip tape over twice and seal them with a grip sleeve.

Use two pairs of wrenches to tighten the tape loc a quarter turn past hand tight to prevent leaks.

assembling the system

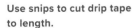 Once you've finished your irrigation map, use it to make a shopping list for your system. Drip and overhead watering supplies may be available at local garden centers and nurseries, and are readily available online. Before buying any irrigation supplies, make sure that you can find a full set of compatible materials. Remember: *Different brands of irrigation materials may not fit together, even if the diameter of the tubes and fittings are the same.* There is nothing more frustrating than trying to fit two sets of parts together that almost fit, but not quite.

THE RIGHT TOOLS FOR THE JOB

Here are the tools you'll need to quickly and easily set up and maintain your drip irrigation system.

+ **Hole punch.** This is a tool specially designed to create the right diameter hole into tubing, so that you can insert drip lines or emitters. Buy one that's designed for use with your particular system. Some are fancy and ergonomic; some are simply a plastic handle with a metal tip. They all do the same thing.

+ **Cutting tool.** Use a pair of scissors, a knife, tin snips, or any other cutting tool to cut pieces of poly tubing and drip lines to length. Don't use your gardening hand pruners or they'll get really dull. PVC pipe can be cut with a standard hacksaw, or use a specific PVC pipe cutter to make the job easier.

+ **Wrenches.** Keep two pairs of wrenches handy. Some drip fittings need to be tightened beyond "hand tight." Wrenches are also handy to have in case the water spigot itself has a leak. Channel locks, locking pliers, or crescent wrenches work well.

RAIN GAUGE OR SOIL MOISTURE METER

You might consider installing an electronic rain gauge or soil moisture meter as part of your irrigation system. A rain gauge monitors the relative air humidity. When humidity goes up to a certain level, the gauge turns the irrigation system off temporarily, because it uses this metric as an indicator of rainfall. Rain gauges can be effective for water conservation but can also misinterpret weather situations. If you live in an area with high humidity levels (even without rain), keep an eye on the rain gauge and irrigation system to make sure everything is functioning properly. Soil moisture meters attempt to read the actual moisture level in the soil, but are not always reliable. Both of these tools can save a significant amount of water, but it's still best practice to keep a vigilant grower's eye on your garden's water needs.

+ **Teflon tape.** Also called plumber's tape, this white, stretchy tape can be wrapped around threaded fittings. It is not essential, but it can really help tighten up the system and keep it drip-free, especially when used on threaded metal fittings.

+ **Rags.** A few small rags can be incredibly helpful to clean dirt off irrigation pieces that you dropped in the soil or otherwise got dirty. Remember, keeping soil out of the system will prevent clogs.

+ **Shovel and hand trowel.** Both of these tools might be necessary if you plan to bury any of the tubing or if you need to move soil out of the way when adding tubing to an existing bed.

+ **Drill.** Optional tool that you might use if you want to attach the tubing to the side of the house or side of a bed to help keep it in place. You can use ½- or ¾-inch pipe clamps to secure tubing to any workable surface.

When you have everything on hand, begin assembling the irrigation system at the water spigot. Take your time and fit the pieces together tightly. Take care when cutting pieces of mainline tubing to ensure they are the right length. Once fully assembled, remove the end fittings of your drip lines or the heads on your sprinklers, and run water through the system to flush out any dirt that may have gotten in during setup and any tiny pieces of plastic that traveled from the factory. After running the system for a few minutes, turn off the water and seal up the ends of every line or install heads on your sprinklers. Run the system once again, carefully examining every piece, fitting, and drip line or sprinkler for leaks. Take the time to tighten any dripping fittings, add couplers, or replace any lines that you may have inadvertently cut while building the system.

WINTERIZING

Most irrigation lines are not under pressure between watering cycles, so tubing, sprinklers, and drip lines tend to empty themselves of water after each session. At the end of the season, if you've used PVC for any part of your irrigation system, make sure these components are completely free of water, as they crack easily. When cold weather is approaching, turn off the water spigot and remove the Y-valve, timer, pressure regulator, and filter and put them inside for the winter. If possible, turn off the water supply to the spigot (unless it is a frost-free spigot) and/or cover the spigot with a thermal cover.

IRRIGATION REPAIR KIT

Keeping a few repair pieces in your toolbox can make maintenance of an irrigation system a breeze. We keep a small bag of irrigation fittings with our gardening tools, to make minor repairs a cinch. After all, the easiest time to notice leaks is while you are working in the beds, so keeping materials on hand will make it easy to keep the system in working order.

If you have an irrigation system, you'll find the following items essential.

- × **Teflon tape.** Keeping a roll handy allows you to quickly seal up small drips from threaded fittings (see facing page).

The components of an irrigation repair kit, from left to right: Hose nozzle with extra washers, mainline tubing couplers, extra bubblers, extra microsprinkler heads, sod stakes, Teflon tape, 9-volt battery, goof plugs, irrigation hole punch.

- × **A spare hose nozzle.** One of the most commonly broken tools in the garden.

- × **Irrigation couplers.** For quickly patching up holes in the irrigation system.

- × **Hole punch.** For moving around drip lines or any other system repairs.

- × **Goof plugs.** For plugging small irrigation holes in header tubing, or to close up the old holes when you move an irrigation line.

- × **An extra battery.** In case your irrigation timer or any other battery-operated tool goes awry. Keep batteries in a plastic bag or other weatherproof container.

- × **Extra sprinkler nozzles/heads.** It's good to have these on hand to make quick in-the-field repairs.

- × **Sod staples.** We like to use short, U-shaped sod staples to keep irrigation lines in place. Keep a few extras around to help pin down any misbehaving parts.

Open up the ends of the mainline tubing to allow excess water to drip out. After leaving the system open for a few minutes, seal the ends back up to prevent dirt or insects from getting inside. Polyethylene tubing is somewhat flexible, so it's possible to leave most types of irrigation lines in place through winter. If you live in a region with extremely cold winters, consider bringing irrigation tubing into a garage or other protected structure to minimize plastic deterioration.

If your irrigation system has components that don't drain easily, which may be a problem with permanent sprinkler systems, consider using an air compressor to blow the water out of it at the onset of winter. Water left in small fittings and emitters can freeze and cause damage, so taking the time to properly drain your system will pay off in reduced repairs and maintenance in spring.

5

EXTEND & EXPAND THE HARVEST

↓

HARVEST LONGER & MORE OFTEN & STORE CROPS PROPERLY.

↓

Maximizing production from each plant is not the only way to get more food from your garden. You can also use tools and techniques to plant earlier in spring, harvest later in fall, and maybe grow food throughout winter. Proper technique and timely harvesting let you use more of your crop and enjoy it at peak quality and flavor. And proper storage methods keep your harvest in top shape long after you've brought the produce in from the garden.

CHAPTER FOURTEEN

EXTENDING THE GROWING SEASON

Season extension is the art and science of using protective covers like cold frames, low tunnels, or a greenhouse to grow and harvest crops earlier in spring, through hot summer spells, and later into fall and winter. Along with storage crops and succession planting, season extension can help you maintain a steady, consistent supply of produce from your garden year-round.

There are simple or complex systems for extending the growing season, depending on your needs. We recommend starting small and scaling up over time as you find the practices and materials that work best in your climate and for your preferred crops.

Structures like a greenhouse can extend the growing season for crops like radishes into spring and fall.

growing in every season

Although weather patterns and seasonality will dictate when the majority of your growing takes place, it's almost always possible to extend and expand your gardening through the entire year. Off-season productivity will be limited and plant growth slower, but surprisingly large jumps in yearly harvest totals can be accomplished with some simple season-extending techniques and good planning.

WARMING UP & COOLING DOWN IN SPRING & SUMMER

In cool spring temperatures, almost any crop will germinate and grow faster when protected by a low tunnel or greenhouse. You may be able to set out plantings of cold-tolerant crops such as peas, cooking greens, salad greens, radishes, turnips, and carrots as early as February or March, and use a row cover as additional protection to help

Slitted plastic row cover can help increase temperatures for heat-loving crops like cucumbers.

them along through the cold weather. You can even sneak frost-sensitive crops such as tomatoes, cucumbers, and summer squash outside as much as a month before your average last frost date.

When it starts to warm up, you can use shade cloth to stretch the growing season of cool-weather crops and protect summer crops from abnormally hot weather. By swapping out your row cover with a shade cloth, you can keep salad greens and brassicas cool, producing tasty harvests even at the peak of summer. You can also use overhead irrigation on crops during hot summer months to help dissipate the heat from their leaves, allowing them to hold longer in the field.

In areas with cool summers, tunnels can provide the extra heat needed for heat lovers like tomatoes, peppers, basil, and eggplant. In this type of setting, a greenhouse can easily quadruple yields of these crops.

HEAT-LOVING CROPS

Greenhouses can help increase yields of heat-loving crops even in climates with relatively hot summers. Not only does the covered structure allow you to plant the crops earlier in spring, but it also keeps precipitation off of the leaves of the plant and the soil around it. This can reduce the

risk of many diseases and give you more control of the postharvest quality of produce.

For example, professional growers all over the United States often choose to plant tomatoes in a greenhouse. This allows the plants to mature earlier in the season and produce later into fall. Because many tomato varieties have indeterminate growth habits, they'll continue to produce fruit as long as soil fertility and weather permit. Cover can also reduce the incidence of early blight and increase marketable yields by reducing cracking and splitting during rain events. Other examples of crops that are commonly grown in greenhouses to increase yield and quality are peppers, eggplant, trellised cucumbers, basil, and raspberries. Sprawling crops like melons and sweet potatoes also produce well in greenhouse culture, but are not as commonly grown there because they require a large amount of space per plant.

It's important to remember that greenhouse culture can reduce some diseases, but will create an environment that favors others. Problematic greenhouse diseases include botrytis (gray mold), white and leaf molds, and powdery mildew. The best way to combat disease is to maintain adequate ventilation in your greenhouse and to prune your crops to promote air circulation.

EXTENDING THE HARVEST IN FALL & WINTER

Fall season extension is a little more complicated. The limiting factors to fall and winter cropping are cold temperatures, inclement weather, and lack of light. You can address the temperature and weather issues by selecting crops that grow well (and sometimes actually taste better) when grown in cold conditions, and by using protective structures. You can address the light issue by growing crops to maturity before day length begins to decrease.

Depending on your latitude, daylight hours may decrease to the point where plant growth slows down to a crawl. In northern regions, daylight will drop to less than 10 hours sometime in October or November, making it difficult for crops to grow without supplemental lighting. In these situations, you're not really *growing* crops in winter; you're *storing* them in the soil.

You have two goals in this scenario. The first is to grow appropriate crops to a harvestable stage before light levels drop, so that you can harvest them throughout the cold months. The other is to germinate seeds in fall, which will go dormant for winter and wait until day length increases, for an early-spring harvest. Skilled growers throughout the United States, including frigidly cold northern areas in Maine and areas without much winter daylight in the Pacific Northwest, use these techniques and season-extension devices to harvest through winter.

CHOOSING COOL-WEATHER CROPS

Planting cool-weather crops is a great way to extend your growing season. Many crops can tolerate light frost, and some can withstand hard frosts and snow.

Cool-weather crops grow well in early spring, fall, and winter. Winter success depends on your climate and how much protection you give them. The temperatures and planting dates we list on

NITRATES IN WINTER-HARVESTED GREENS

Some studies have shown that greens grown during periods of low temperatures and low light levels can take up nitrates from the soil into their tissue. Consumption of nitrates is believed to carry health risks. Best practice is to harvest greens grown during late fall and midwinter in the afternoon or early evening, and preferably on a sunny day. Exposure to even a short period of sunlight drastically reduces the nitrate levels in the plants' leaves.

pages 230–231 are general and based on observation, rather scientific research, so your experience may vary from ours. We suggest reading further (see Resources, page 292) and experimenting on your own to fine-tune the best planting dates for your climate.

Cool-season crops like broccoli can continue to produce harvests like these side shoots for months after the main head is cut.

Our favorite way to take advantage of cool-weather crops is to grow a variety of baby leaf crops and combine them into unique salad mixes. For early-spring production, sow these crops in a greenhouse as early as you can work your soil, or outdoors under a low tunnel. For fall production, sow them in late August through September.

In most sections of the northern United States, you can maintain a harvest into winter using a greenhouse or a low tunnel with heavy polyester row cover (or a polyester and plastic combination). To overwinter and harvest in the coldest months, you might need a low tunnel and greenhouse combination.

Because establishing new plantings is difficult, if not impossible, in late fall and winter, using "cut-and-come-again" culture works best for many of these greens. When the baby leaves are ready to harvest, cut them off cleanly about an inch above the soil level. If tiny new leaves are starting to emerge from the base of the plant, cut above them to maintain the quality of the next cutting. After harvesting, the tiny new leaves will slowly grow into a harvestable size. Depending on the crop, soil, and weather conditions, you may get anywhere from two to four cuttings from a single planting.

COLD-HARDY SALAD CROPS

× **Baby beet leaves.** Can tolerate temperatures as low as 20°F (−7°C) when leaves are young. Adds color and flavor to salads, or can be sautéed.

× **Baby kale.** Cold tolerant to around 15°F (−9°C). 'Red Russian' baby leaf is especially excellent in salads.

× **Baby Swiss chard.** Very cold tolerant when leaves are young. Try 'Rainbow' or 'Bright Lights' for beautiful colors in your salad mix. Full-grown chard tolerates frost well, but not repeated freezing.

× **Claytonia.** Cold tolerant to well below 20°F (−7°C). This green adds great weight and texture to salad mixes. It maintains its raw eating quality even when flowering. Thin plants to 4 inches apart for best growth.

× **Curly cress/peppercress.** Tolerant of frosts and freezing. Good for multiple cuttings, it has great flavor and is easier to grow than watercress.

× **Greens mixes.** Cold tolerance depends on which species are included. Greens or "braising" mixes make it easy to grow a range of hardy greens in a single planting, saving valuable space. Avoid greens mixes that include lettuces, as these will have different growth rates and cold tolerance than the other species in the mix.

× **Mâche.** Tolerates temperatures as low as 15°F (−9°C), but it can be a little finicky to grow. Some people love its unique flavor, and others find it off-putting. It's usually harvested as a whole plant; it doesn't regrow well for cut-and-come-again culture.

- × **Minutina.** Tolerates light frost. It can get tough and bolt quickly after the first cutting. It has an interesting flavor and texture.
- × **Mizuna.** Recovers well from temperatures as low 15°F (−9°C). A staple winter green, its mild, pak choi–like flavor is excellent in salad mixes, and it can also be grown larger for braising or stir-frying. It has strong regrowth for cut-and-come-again culture.
- × **Mustard greens.** Tolerates temperatures down to 25°F (−4°C), depending on the variety. Adds a great kick to salad mixes as a baby leaf, and can also be grown full size for sautéing. There are a variety of colors and shapes available.
- × **Pak choi/tatsoi/other Asian greens.** Tolerates temperatures down to 25°F (−4°C), depending on the variety. Winter and early-spring plantings can bolt quickly when temperatures warm up. They're excellent in salads as a baby leaf, or can be grown to maturity for cooking.
- × **Spinach.** Very cold tolerant. We've seen it recover from temperatures close to 10°F (−12°C). You can use cut-and-come-again culture when leaves are young for raw eating, or you can pick individual larger leaves for cooking to extend the harvest over a longer period.

MODERATELY COLD-HARDY SALAD CROPS

- × **Arugula.** Handles light frosts well, but starts to deteriorate with repeated hard frosts. This spicy green is mellow and delicious when grown in cooler temperatures. Can be cut several times, but quality is best at first cutting.
- × **Baby lettuce mix.** Lettuce tolerates light frosts well when leaves are at baby size, but it loses quality when temperatures get as low as 25°F (−4°C). Heads of lettuce grow well in cool spring and fall temperatures but don't tolerate below-freezing temperatures well.

OTHER GREAT COOL-WEATHER CROPS

- × **Baby-size turnips.** These tolerate frost, but don't let the roots freeze. Varieties such as 'Tokyo Cross' and 'Hakurei' size up quickly in a short period (they're mature when at golf-ball size). In spring, sow as early as you can work the soil; sow your fall plantings into early September if using a low tunnel or greenhouse.
- × **Carrots.** These are sweetest when they are grown in cool weather and spend some time at temperatures near freezing when mature. Their greens can tolerate below-freezing temperatures, but don't let the roots actually freeze or they'll become fibrous and tough, or turn to mush. To prevent roots from freezing, cover overwintering plantings with 6 to 18 inches of straw or another easily removable protective mulch. In spring, sow as early as you can work the soil; sow your last outdoor plantings in late June to mid-July, or even into early August if using short-day varieties and a low tunnel or greenhouse.
- × **Radishes.** These tolerate frost well, but lose quality if the bulbs fully freeze. In spring, sow as early as you can work the soil; sow your fall plantings into early September if using a low tunnel or greenhouse.
- × **Scallions.** Extremely cold tolerant, and can be grown outdoors without additional protection in many climates. In spring, sow as early as you can work the soil; sow your fall plantings into early September if using a low tunnel or greenhouse.

Low tunnels are inexpensive and easy to set up.

choosing a structure

↓   Protective structures are the key to successful season extension. They fall into two general categories: short structures, such as low tunnels and cold frames; and tall structures, like greenhouses. Short structures have the advantage of being less expensive and easier to install, while tall structures allow for easier crop management and a wider range of crops. Tall structures also provide room for you to set up a lounge chair and take a nap on sunny spring afternoons.

LOW TUNNELS

Low tunnels, also known as row covers, are structures that are not tall enough to walk into. They are often made of hoops or arches that support fabric or plastic covers to protect plants. To manage and harvest the plantings underneath, the covers must be removed. Low tunnels are inexpensive and easy to set up, but because they're low, you can't grow tall crops to maturity under them.

For simplicity of construction and minimal cost input, low tunnels are often built 2 to 4 feet tall. A shorter cover will hamper the growth of a maturing kale, peppers, or summer squash, but will be perfect for salad greens and carrots. A low tunnel can easily be constructed to a height of up to 4 feet, which is appropriate for crops like peppers, tomatillos, and broccoli. If you feel the structure needs to be any taller than 4 feet, consider upgrading to a greenhouse (see page 235).

HOW TO CONSTRUCT A LOW TUNNEL

It's pretty easy to improvise a structure for a low tunnel, or you can purchase materials from a nursery or seed company (see Resources, page 292). Common materials for hoops are #9 or #10 gauge wire, or PVC pipe bent over rebar pins pounded into the ground. When using rebar, we like to use 2-foot lengths of ½-inch rebar for pins and 10-foot lengths of electrical conduit cut to size to fit over the bed. Another easy option is to bend a section of wire-mesh fencing into a C shape and attach your covering material to the wire with zip ties, twist ties, twine, or bungee cords. Low tunnel structures, especially taller ones, will be more stable with a top purlin.

Low tunnels can be covered with a variety of materials, including:

Spun-bonded polyester. This fabric provides some heat retention and wind resistance but is also breathable and lets precipitation pass through it. It doesn't create a greenhouse effect like clear plastic does, so you don't have to remove it when the sun comes out. You can purchase polyester row cover (Reemay and Agribon are common brand names) in different weights—heavier weights provide the most protection from the elements; lighter weights are more breathable and are best for pest exclusion (see page 155). It will let anywhere from 30 to 90 percent of the sun's rays pass through, depending on the weight of the fabric.

Greenhouse or UV-resistant plastic film. Using clear plastic to cover your hoops creates a greenhouse effect, so you get maximum temperature gain in the bed. However, it can fry crops on a sunny day—even in the middle of winter. You'll need to be prepared to open it partially or remove it for ventilation during sunny periods. It doesn't breathe and doesn't allow precipitation to pass through, so you'll also need a way to water the crops underneath. Greenhouse plastic lets 80 to 90 percent of sunlight pass through. Only

KEEPING IT TOGETHER

You can use a variety of techniques to hold row covers in place:

Attach the covering to the structure. Purchase clamps or specially made clips to help attach material to your hoops or other row cover structure.

Hold down the bottom edges. Most home growers use sandbags, rocks, or bricks to weigh down the edges of their fabric. This works well, but involves some labor to move them around and store them when they aren't in use. You can also use soil to bury the edge of the row cover, especially if you don't mind doing some extra digging. This method works particularly well if excluding pests is your main goal because you can effectively seal the entire edge of the cover with soil. Some growers use stakes or landscape staples to hold the covers in place, but this can destroy the fabric very quickly.

Close the ends. Seal the ends of the row covers by bunching the fabric or plastic together and holding it in place with weights or soil.

When purchasing materials for a low tunnel, make sure to get hoops and a row cover that are wide enough to cover your beds adequately. Account for the

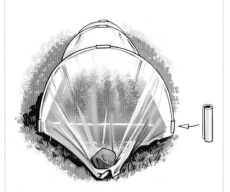

extra length needed to arch over the bed and the extra row cover needed to place weights on or secure with soil. For example, an 83-inch-wide row cover is just about the right size to span a low hoop over a 4-foot bed.

purchase plastic designed for crop production because it is designed to withstand exposure to the sun. Avoid using clear plastic tarps or painter's plastic purchased from a hardware store; light transmission may be poor, and the plastic will break down very quickly because it's not UV-resistant.

Slitted plastic film. This UV-resistant plastic film contains slits or holes for ventilation and to reduce temperature spikes in sunny weather. The slits remain closed when the plastic is cool, but when it heats up and expands, the slits open and allow air to pass through. You still need to water underneath this type of cover, and will need to open it up to ventilate during long sunny periods; but it's much more versatile and requires less management than a solid plastic row cover. Because it's less insulating than solid plastic, slitted plastic is more commonly used to warm up spring crops than to protect overwintering crops.

Cold frames can be attached to virtually any garden bed.

COLD FRAMES

Cold frames serve the same purpose as low tunnels but have a sturdier structure built from wood or metal. Because of the sturdy frame, you can use heavier covering materials such as double-wall polycarbonate sheeting or salvaged windows. Try to find plastic or tempered glass windows; old glass can break and leave shards in the soil. Greenhouse plastic or spun-bonded polyester also work.

The advantage of cold frames is that they are less likely to blow away than low tunnels and have more insulation value. Also, you don't have to assemble or disassemble them—you can just lift up the whole structure and set it over a garden bed. The drawback is that they don't cover a lot of square footage in the garden relative to the time and expense of building them. Also, they can be difficult to move around (most take two people), and you have to find somewhere to store them when they're not in use. We find that they're most useful for protecting small amounts of space-efficient salad greens, or protecting a few flats of transplants while they're growing.

Cold frames can also be installed as permanent structures attached to your raised beds. You can take off the covers during summer or swap in shade fabric during periods of extreme heat.

USING LOW TUNNELS & COLD FRAMES

Almost any crop can be covered with a low tunnel or cold frame in spring to enable earlier planting and harvest. For frost-tolerant crops, sow or transplant as soon as you're able to work the soil. Water in, then cover with plastic, slitted plastic, or spun-bonded polyester row cover. For frost-sensitive crops, transplant from 2 to 6 weeks earlier than suggested by the Planting Dates Worksheet on page 274. The additional thermal and wind protection provided by the cover will accelerate harvest by 1 to 5 weeks.

Exact timings will depend on your local climate/microclimate and your tolerance for risk. We know a skilled professional grower in Pennsylvania who sets out his first planting of tomatoes under row cover on March 15, a full 2 months earlier than is recommended for his climate. If spring works in his favor, he's the first in his market to show up with tomatoes. If it's unusually cold, he might lose the whole planting and have to set out another one. He feels it's worth the risk because early tomatoes are prized by his customers.

With plastic and slitted plastic covers, you'll need to closely monitor temperature under the cover and be prepared to remove it when necessary (consult the Planting Dates Worksheet on page 274 for ideal growing temperatures for each crop). You also need to remove the covers for watering if you don't have drip irrigation installed underneath. With spun-bonded polyester covers, you don't need to worry about removing them, except if the plants are flowering and need insect pollination, to harvest, or if the plants need more room to grow. This may be the case for tall crops like trellised tomatoes.

EXTENDING YOUR HARVEST IN FALL

Transplant or direct seed your fall crops, wait until temperatures and/or light levels start to drop, and then cover them with a floating row cover of your choice. For example, you might sow spinach on September 1 for fall harvest. If your early-September temperatures are still 70°F (21°C) or above, wait until late September or October to install row cover.

We like to use heavy row cover for this application, because our primary concern is protection from wind and cold, not light transmission. In many regions, once you're into October, there's not enough daylight for plants to do much growing anyway.

You can also use multiple coverings on your low tunnels. In early fall, start with a spun-bonded polyester row cover over your low tunnel. When temperature and light levels drop in late fall or early winter, add a greenhouse plastic cover on top of the fabric.

A greenhouse with an inner cover directly over each bed helps produce crops weeks earlier in spring than under outdoor conditions.

GREENHOUSES/HIGH TUNNELS/HOOPHOUSES

"Greenhouse," "high tunnel," and "hoophouse" are essentially interchangeable terms—they all describe an arched or gabled structure that is tall enough to walk into. Some growers use the term "greenhouse" if the structure has a heat source and "high tunnel" if it's unheated, but this isn't universally adopted.

Greenhouses are more complicated and costly to set up than low tunnels, but they're much easier to work in. They also allow you to grow tall crops, such as tomatoes and cucumbers, to maturity. An added benefit of using a greenhouse is that you can keep precipitation off the

A small greenhouse and row covers can dramatically increase the yield of a garden.

A simple structure like a caterpillar tunnel is cost-effective and easy to set up.

leaves of disease-prone crops, which can help extend their season dramatically.

Usually, greenhouses have doors or windows on the ends for access and ventilation, and are outfitted with some way to open or roll up the side walls for maximum ventilation during hot days. If you have access to electricity, greenhouses can be fitted with automatic ventilation shutters and fans and can be heated for further season extension.

Heating a greenhouse does have a high environmental cost in terms of fuel use and carbon emission, so we recommend only doing so when absolutely necessary. Growing transplants may be a good reason to provide supplemental heat, but winter harvest extension may not. You can still make amazing leaps in season extension and improved yields using an unheated or minimally heated greenhouse.

THE STRUCTURE

The hoops and arches that support a greenhouse are usually made from bent steel pipe. You can purchase these premade, or you can bend them yourself. Johnny's Selected Seeds sells a great bender for small-scale growers, which is often used for construction of "caterpillar" tunnels (so-called because their segmented structure

makes them look like enormous insects). These low-cost greenhouses are constructed from materials that are easy to source. Most components are pieces designed for chain-link fencing. See Resources (page 292) for more information.

Small greenhouses may only have hoops or arches without additional structure. As they get larger, purlins, ridge poles, corner braces, and trusses become necessary. You can construct greenhouse end walls from steel pipe with customized fittings, or you can make them using standard framing lumber. A "gothic" arch shape is usually more expensive, but is stronger and handles snow loads better. Older greenhouses and smaller greenhouse kits are often made with a more traditional wood or metal frame with a gabled roof. This works great for kits, but metal hoops/arches are more cost effective for larger greenhouses.

FINDING THE RIGHT GREENHOUSE

There are virtually endless brands and materials to choose from, and a wide array of specific hardware pieces that might be appropriate for your greenhouse. Identify a local or regional supplier and ask them to help you design the structure and make sure you have the appropriate fittings. If all you need is a 6- by 6-foot or 8- by 8-foot structure, consider buying a kit. This saves you the trouble of figuring out all the fittings and parts needed to assemble a greenhouse using metal hoops.

GREENHOUSE COVERINGS

Greenhouses are almost always covered with UV-resistant plastic film or acrylic panels. Old-school greenhouses and newer high-end ones use glass. Glass greenhouses are beautiful and trap heat well, but they are expensive, break easily, and can be difficult to repair. Double-wall polycarbonate panels have more insulating value than plastic sheeting but are quite expensive. These panels are often used to construct endwalls.

If you have a greenhouse covered with plastic film, you can increase the insulation value by putting two layers of plastic on, and inflating the air space in between the layers with a specially designed electric blower.

To provide maximum protection in cold northern climates, you can use low tunnels inside an unheated greenhouse to grow cold-hardy greens year-round. These are usually called inner covers and can help keep crops alive even when outdoor temperatures fall to 10°F (−12°C) and below!

SITE SELECTION & ORIENTATION

When choosing a site for a greenhouse, consider its orientation. For summer production of heat-loving crops, some growers prefer a north–south orientation of the long axis because it allows for more even light distribution on your plants. In our experience in the northeastern and northwestern United States, we haven't observed a big difference between east–west and north–south orientation for summer crops. For fall and winter production of greens, many growers prefer an east–west orientation for maximum light intensity in winter. Our experience has found this to be very effective.

CHALLENGES OF GREENHOUSE PRODUCTION

One issue with greenhouses is that the soil under them is never exposed to precipitation and extreme temperatures. This can lead to nutrient imbalances (high salt levels in particular) and a buildup of pest and disease problems. Careful ventilation, fertilization, and irrigation will help deal with these issues, but some growers are pioneering techniques using movable greenhouses to expose the soil to weather and allow for longer crop rotations. These movable structures usually slide back and forth on skids or on tracks with rollers, and are moved by hand or with a small tractor. Alternatively, it's possible to disassemble and move a greenhouse if problems arise.

In the United States, we recommend the following guidelines.

+ If you're north of 40 degrees north latitude, orient your greenhouse east–west for maximum light intensity.

+ If you're south of 40 degrees north latitude, light intensity is much less of an issue. Orient your greenhouse north–south for best light distribution.

INTERIOR LAYOUT

You have many options to consider when laying out beds in a greenhouse. The simplest technique is to set up beds that run the length of the structure. If you grow trellised crops, such as cucumbers or tomatoes, an added benefit to this layout is that you can use purlins to hang trellis strings from. If your greenhouse has sloping sidewalls, you'll probably want narrow beds on the outside edges of the structure, so your path ends up where the greenhouse is tall enough to walk without stooping.

Keeping a thermometer in your greenhouse is a good way to monitor the temperature inside your structure. If you don't have automatic venting, we recommend using a smart thermostat that can alert your phone whenever the greenhouse needs to be opened or closed.

If you have vertical sidewalls, awkward spaces along the wall may not be an issue. The perimeter of a greenhouse is usually the coldest space, so there are arguments for putting a path along the outside edge. In our experience with garden design and production growing, we feel that good access is the most important consideration—if it's hard to access beds because you hit your head on the greenhouse frame, you won't take as good of care of them.

Another option that maximizes plantable square footage in the greenhouse is using comb-shaped or keyhole beds. These are a little more

LAYING OUT BEDS IN A GREENHOUSE

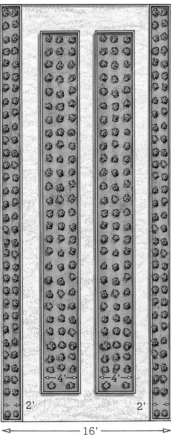

This 16'-wide greenhouse is laid out into rectangular beds for easy planting in straight rows. Note the narrower beds on the outer edges for ease of access.

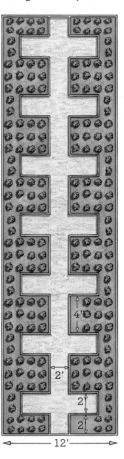

This 12'-wide greenhouse is laid out into comb-shaped or keyhole beds to maximize usable square footage in the space.

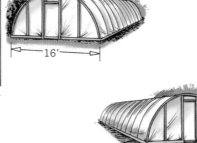

complicated to lay out initially and are a little harder to manage if using low tunnels within a greenhouse, but they allow you to plant more of the protected square footage of the greenhouse.

LIGHTING & HEATING

For most season-extension purposes, the added expense and complication of adding a light and heat source to the structure will be prohibitive. However, there are applications in which this extra work will be worthwhile, especially when producing young transplants. Please see pages 170 and 173 to learn about adding supplemental lighting and heating to your greenhouse.

USING GREENHOUSES/ HIGH TUNNELS/HOOPHOUSES

For spring and fall planting, greenhouses are managed just like low tunnels. The major difference is that the greenhouse is still standing during the main part of the growing season, so you can use it to grow heat-loving crops such as basil, tomatoes, cucumbers, or peppers.

A typical strategy is to begin establishing fall crops in late August and September. These crops will be harvested into late fall/early winter, or may overwinter depending on crop choice, climate, and if you use an additional low tunnel over the crop. In spring, continue to harvest overwintered crops and clear as they finish producing or as they bolt. Plant spring and summer crops as soon as winter crops begin to decline or you no longer want them. If necessary, you can relay plant warm-season crops into the tunnel before the overwintered crops are completely removed.

When the warm-season crops have finished producing in fall, start planting fall greens again. Many warm-season crops are still producing in early September when you need to begin fall planting, so you'll have to prioritize either maximizing summer yields or removing a portion of these crops early to make room for the next season.

COVER CROPPING OR FALLOW PERIODS IN A GREENHOUSE?

Some growers who are focused on fall and winter crops may not use their greenhouse for vegetable production in summer. Instead, they'll plant a short-season cover crop like buckwheat or mustard greens (see page 89 for more information on cover cropping). This is a good way to manage soil quality and disease issues. Some growers who are primarily focused on growing warm-season crops in their greenhouse might leave it fallow in winter, or even remove the covering to expose the soil to cold temperatures and precipitation. However, because the greenhouse is such a valuable production space, most growers use it for growing crops year-round and address issues that arise using other organic methods.

VENTILATION & DAILY MANAGEMENT

Ventilation is crucial for moderating the temperature and minimizing disease in a greenhouse. If you don't have automatic vents and fans in your structure, you'll need to be available on a daily basis to open it during sunny periods to prevent

Ventilation is key to the success of any season-extending structure. Roll-up sides on a greenhouse can provide cross ventilation.

Even self-ventilating slitted row cover may need to be taken off during hot weather.

the temperature from getting high enough to damage your crops.

Hang a thermometer in your greenhouse so you can easily monitor the interior temperature. Consult the Planting Dates Worksheet on page 274 for guidance on what temperature is best for each crop. If you're growing peppers and tomatoes in spring, you might wait until the temperature is above 75°F (24°C) before venting the greenhouse. If you're growing spinach or salad greens, you might vent it at 65°F (18°C).

To minimize disease, consider opening the greenhouse on cloudy days for a short period at midday even if interior temperatures are still low. If you have electricity in your greenhouse, fans are great to help with air circulation in spring and summer. Wind can damage crops when it's cold/below freezing, so we don't set up fans until temperatures are above 50°F (10°C). When temperatures are higher, fully open the ends and sides for maximum airflow. Roll-up sides are helpful for ventilation and are especially critical

for structures that are more than twice as long as they are wide. In fall and winter, close up the greenhouse an hour or so before you lose sunlight, to trap warm air before outdoor temperatures start to drop.

USE INNER COVERS FOR EXTRA PROTECTION

If you're going for winter-long production of greens in colder climates, plan to use low tunnels (inner covers) inside your greenhouse. Start by establishing the greens in your greenhouse as you normally would. When outdoor temperatures start to drop below freezing, set up a low tunnel over the greens with a heavyweight fabric or greenhouse plastic row cover. If it's sunny, remove the row cover after the temperature in the greenhouse is above freezing. Replace it in the late afternoon before the greenhouse temperature drops below freezing. If it's cloudy and the temperature stays below freezing all day in the greenhouse, leave the cover on.

cooling in summer can increase yields

In hot climates, and even in temperate climates that are prone to heat waves, protection from very high temperatures may be necessary to keep crops thriving and productive. Most of the cool-weather crops mentioned on page 230 can be grown longer into the hot season when given a bit of shade.

Even heat-loving crops can suffer when temperatures jump above 90°F (32°C) for extended periods of time. High temperatures can result in blossom-end rot and dropped blossoms in many fruiting crops. In extreme heat, tomatoes, peppers, and beans can stop setting fruit altogether. A heat wave that brings temperatures into the triple digits can stress out even the hardiest vegetable crops. Adapting your irrigation regime to help cool crops and utilizing your semishady areas

can help stretch your garden season to the max. You may also consider covering your season-extending structure with shade cloth.

USING SHADE CLOTH/COVER

Shade cloth can reduce sunlight penetration and keep plants and soil cool. Covered with a layer of shade cloth, almost any low tunnel, cold frame, or greenhouse structure can help extend the summer growing time for cool-loving crops like lettuce and salad greens. If you're covering a greenhouse or other more permanent structure, you can drape a layer of shade cloth over the top of the plastic to provide relief from direct sunlight. In these situations, be extra conscientious about ventilation.

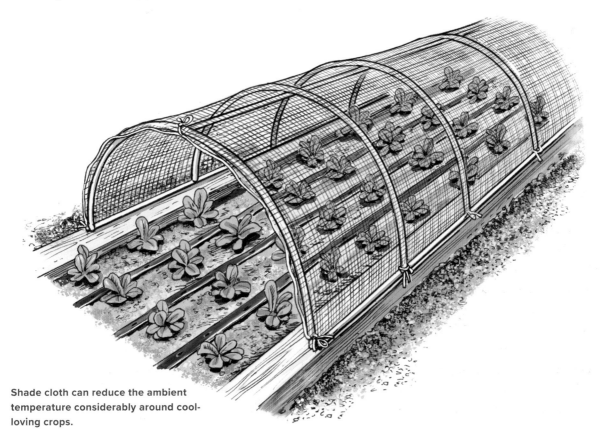

Shade cloth can reduce the ambient temperature considerably around cool-loving crops.

Shade fabric covers make it possible to grow heat-sensitive crops during the peak of summer.

ADAPTING YOUR IRRIGATION REGIME OR USING SHADE CLOTH TO HELP COOL CROPS CAN HELP STRETCH YOUR GARDEN SEASON TO THE MAX.

PLANTING IN PARTIAL SHADE

Try to use the portions of your garden that receive a little less sunlight for summer plantings of heat-sensitive crops. You may even consider building a special bed in a semishady location as a summer greens bed, taking advantage of the space when sunlight is ample enough to allow the plants to grow but not so excessive that they become heat stressed. The east side of a house might be a good location for such a bed, allowing the plants to receive morning light but find reprieve by midafternoon when the sun is at its peak.

OVERHEAD IRRIGATION TO COOL CROPS

Even if you've decided to use drip irrigation for your garden-watering needs, a quick and easy overhead irrigation setup can work wonders for cooling crops in summer. Sprinkling crops from overhead will cool down the leaves, stems, and fruits of the plants, as well as the soil around them. You won't want to turn on sprinklers in the middle of the day, but in very hot weather you can overhead irrigate in the morning and again in the late afternoon. The afternoon watering will help cool down the plants for the night but should be done early enough in the day that some of the moisture has time to dissipate from the leaves before dark. We have seen growers harvest perfect heads of lettuce in the middle of July by employing a regular schedule of watering in this way.

In hot climates or during abnormally scorching summers, shade cloth can be worth its weight in lettuce. It's breathable and lets precipitation pass through it. You can select different fabric grades to supply the amount of shade you need. Very generally, 30 percent shade works well for protecting fruiting crops from sunburn in hot areas and for growing lettuce and salad greens in midsummer in the northern regions. Fifty percent shade cloth works well for shade-loving plants, lettuce, and salad greens in hotter climates.

If you live in a particularly hot climate, you may even consider setting up shade cloth over the entire garden in peak season. Using tall stakes, you can build a temporary lean-to over the garden, casting shade with minimal investment and time.

CHAPTER FIFTEEN

TIMELY HARVESTING & SUCCESSFUL STORAGE

Harvesting your crops is a singularly rewarding experience. After all of your planning, preparation, and hard work, you finally reap what you have sown (just like the Little Red Hen). There's much to be appreciated about the growing process itself, but harvesting is what really inspires you to keep going.

The goal of a production-minded vegetable gardener is to grow as much food as possible from the space available. However, this only makes sense if you can make good use of the crops you have grown. Proper harvesting technique, timing, and storage will ensure that the fruits of your labor do not go to waste.

becoming a skilled harvester

↓ Harvesting is an important skill unto itself. When and how you do it can contribute greatly to the eating quality and storage life of your produce. Crops are ready to harvest at different stages depending what part of the plant you are harvesting, and at different times of year based on the crop's planting date and life span.

Through proper planning and management, you've spread the harvest out over as much of the year as possible, but in every garden calendar there is always a peak harvest season. In temperate climates, peak harvest season happens in late summer and fall. In hot climates, it may be later in the year, or in spring. No matter when your climate sets the stage for optimal plant growth, you should be ready for action.

WHEN TO HARVEST

Every crop has a characteristic size or appearance that indicates when it is ready to harvest. Anticipating the approaching harvest and recognizing these signs will ensure that you pick your crops at the peak of their flavor and freshness, and are able to store them for the longest period possible.

ANNUAL VEGETABLES & HERBS

+ **Arugula.** Cut with a knife or scissors about 1 inch above soil level when leaves are desired size (usually 3 to 5 inches).

+ **Basil.** Pinch set of four leaves at the top of the plant early on to encourage growth of side shoots, then pinch four leaf sets heavily from the whole plant about once a week. Conversely, cut the top third of the plant off and strip the leaves.

+ **Beans, edible soy (edamame).** Pick pods after they've plumped out, but before they turn brown and leathery.

+ **Beans, fava (broad).** Pick pods after they've plumped out, but before they turn brown and leathery.

+ **Beans, lima.** See Beans, shell.

+ **Beans, shell.** For fresh shell beans, pick pods after they've plumped out, but before they turn brown and leathery. For dry beans, pick after most of the plant's leaves have fallen and pods feel dry and crunchy. Continue drying in a covered, well-ventilated space if weather is not conducive for field drying.

+ **Beans, snap.** Pick when pods are about the thickness of a pencil (before inner seeds start to swell).

+ **Beets.** Harvest when roots reach desired size (you can harvest them small for "baby" beets).

+ **Broccoli.** Cut main head when "beads" are swollen, but before they separate or open. Continue harvesting side shoots for 1 to 4 weeks after main head is cut.

+ **Broccoli raab.** Cut whole plant at base when buds appear, or pick only buds for extended harvest.

+ **Brussels sprouts.** For earlier and more uniform harvest, cut the loose head at the top of the plant when sprouts are about ½ inch in diameter (late August to early September in many climates). Pluck individual sprouts when fully sized up, or cut entire plant at base if desired. Flavor improves when sprouts experience a few frosts.

+ **Cabbage.** Cut head at base when it feels dense, firm, and well filled out (before the top of the head starts to separate).

+ **Cabbage, Chinese.** Cut heads at base when they reach desired size (Chinese cabbage doesn't feel as dense as standard cabbage).

+ **Carrots.** Dig when roots reach desired size (dig one or two up to check; you can't tell from aboveground). Carrots can be harvested young for "baby" crops, but make sure they have full coloration (very young carrots are pale and don't have fully developed flavor).

+ **Cauliflower.** Cut head when it reaches desired size, but before the curd starts to separate. White color can be maintained by tying outer leaves around head when it's small.

+ **Celeriac.** Harvest when roots reach desired size (3 to 5 inches across).

+ **Celery.** Harvest when stalks reach desired size. Celery can be blanched for milder flavor by wrapping/covering the stems during the growing process (keep the top leaves exposed). Try a half-gallon milk jug with the top cut off for individual plants, or two boards placed on either side of a row of plants.

+ **Chard, Swiss.** Harvest outer, larger leaves (with stem attached) about once a week when they reach desired size. Don't remove more than 30 percent of the leaves at one time for best future harvests.

+ **Cilantro.** Cut with a knife or scissors about 1 inch above soil level when leaves are desired size.

+ **Collards.** See Chard, Swiss.

+ **Corn, sweet.** Harvest when ears feel well filled when squeezed, and when kernels are plump and "milky" (you can peel back the wrapper leaves and puncture a kernel with your thumbnail to check). This usually happens 18 to 24 days after ear silks first show.

+ **Cucumbers.** Pick fruit by hand or with shears when they reach desired size (well before cracking or yellowing appears). Once bearing begins, pick every other day to maintain best quality.

+ **Dill.** For dill weed (foliage), cut with a knife or scissors when leaves reach desired size. When plants are small, cut-and-come-again culture is possible, or individual branches can be harvested on a weekly basis. For seed heads for pickling, harvest seed cluster (called an umbel) when seeds have filled out (you can also use flower clusters before seeds have formed for this purpose). For dill seed, wait until seeds have turned dry and brown on the plant.

+ **Eggplant.** Harvest fruit with shears at desired size. Pick regularly for best production.

+ **Endive.** Cut head at base when it reaches desired size. For milder flavor, blanch the head by tying the outer leaves together over the plant, or covering it entirely with a reused nursery pot or other container 1 week before harvest.

+ **Fennel, bulbing.** Cut bulb at base after it has swollen and reached desired size.

+ **Garlic.** Harvest by lifting entire plant with a spade or trowel when half of the leaves have browned out (half should still be green). See text on curing on page 261 for more information.

+ **Kale.** Harvest outer, larger leaves (with stem attached) about once a week when they reach desired size. Don't remove more than 30 percent of the leaves at one time for best future harvests.

+ **Kohlrabi.** Cut bulb at base after it has swollen and reached desired size (3 to 8 inches in diameter, depending on variety).

+ **Leeks.** Harvest by lifting plant with a spade or trowel when it has reached desired size (approximately 1- to 2-inch-wide stem, depending on variety). Trim roots and leaves with shears for washing and storage.

+ **Lettuce, baby mix.** Cut about 1 inch above soil level when leaves are desired size (usually 3 to 5 inches).

+ **Lettuce, head.** Harvest by cutting head at base.

+ **Mâche.** Cut entire plant about 1 inch above soil level when it's reached desired size (usually 2 to 4 inches).

+ **Melon, cantaloupe, honeydew.** Assessing melon ripeness varies by variety and takes some experience. Most melons are ripe when the fruit shows some yellow coloration and detaches easily from the vine. Honeydew and charentais melons are overripe at this stage and should be cut from the vine using shears. Read as much as possible about the variety you're growing before harvest.

+ **Mustard greens.** Cut about 1 inch above soil level when leaves are desired size (2 to 4 inches for salad eating, larger for cooking). The greens become spicier as size increases.

+ **Okra.** Harvest pods before they get tough (3 to 4 inches long). Harvest every 2 or 3 days for best production and quality.

+ **Onions, bulb.** For baby bulbs and usable greens, harvest at any size when leaves are still green. For best size and storage, wait to harvest until tops have browned and died back. See page 260 for curing information.

+ **Pak choi.** Cut plant at base when it reaches desired size, but before stem elongates and flower stalks appear (4 to 8 inches for baby, 8 to 16 inches full size). Different varieties are mature at different sizes.

+ **Parsley.** Harvest outer leaves (with stem attached) about once a week when they reach desired size. Don't remove more than 30 percent of the leaves at one time for best future harvests.

+ **Parsnips.** For best flavor, dig when roots are fully sized up in mid to late fall, or early the following spring before tops begin to regrow.

+ **Peanut.** Dig peanuts 120 to 160 days after planting, depending on the variety. Loosen a 24-inch-diameter circle around each plant with a spade, then pull the entire plant (including the roots) from the ground, and strip peanuts off the roots. This is best done when the soil is relatively moist.

+ **Peas, shelling.** Harvest when pods are plump and well filled.

+ **Peas, snap.** Harvest when pods are plump and taste sweet (you might need to experiment a bit with different varieties to determine best harvest stage; if you're too early, the peas aren't as sweet; too late and they get tough and stringy).

+ **Peppers, hot.** Harvest when fruit is desired color (green through fully colored red, yellow, or orange). Picking the first full-size peppers at the green stage will improve overall yield of the plant.

+ **Peppers, sweet.** Harvest when fruit is desired color (green through fully colored red, yellow, or orange). Picking the first full-size peppers at the green stage will improve overall yield of the plant.

+ **Potatoes.** For new potatoes, dig at about 8 weeks after planting. For full-size potatoes, dig when foliage is brown and has died back.

+ **Radicchio.** Harvest by cutting head at base.

+ **Radishes.** Pull plants from the ground as soon as stem reaches desired size (varies with variety). Many radishes have a short harvest window and split or become tough or pithy if harvested when overmature.

+ **Rutabagas.** Harvest when roots reach desired size. For best flavor, wait until plant has experienced a few light frosts.

+ **Scallions.** Lift plants from the ground with a trowel or spade when they reach desired size (about twice the thickness of a pencil). Overmature scallions can still be harvested, but are tougher so benefit from cooking like a bulb onion.

- **Spinach.** For baby spinach, harvest like baby mix lettuce. Avoid cutting into the new leaves to maintain quality of the next cutting. For extended harvest and larger leaves, pluck larger outer leaves from plant and leave inner young leaves to size up (this method produces the best regrowth for winter crops).

- **Squash, gourds.** Harvest by cutting stem with shears. Point of harvest varies by variety.

- **Squash, pumpkins.** Harvest by cutting stem with shears when fruit is fully colored.

- **Squash, summer.** Harvest fruit by twisting off or cutting the stem with shears when it reaches desired size. Smaller squash are most tender; larger squash have more flavor. Harvest before outer skin becomes tough.

- **Squash, winter.** Harvest by cutting stem with shears. Point of harvest varies by variety, but many winter squash are fully ripe when they have a distinct "ground spot" where they've been resting on the ground.

- **Sweet potatoes.** Dig roots in fall before soil temperature drops below 50°F (10°C).

- **Tomatillos.** Pluck fruits when they've filled out and their husk starts to split.

- **Tomatoes.** For fully vine-ripened fruit (and maximum flavor), harvest when the tomato is fully colored and is somewhat soft to the touch. Tomatoes ripen well off the vine; for longer shelf life or shipping, harvest when fruit is partially green and harder to the touch.

- **Turnips.** Pull plant from the ground at desired size (golf-ball size for baby, 3 to 4 inches or larger for full-size varieties).

- **Watermelon.** Cut fruit from vine with shears when tiny leaves on tendril close to fruit have turned brown (a ground spot is also a good indicator of ripeness).

PERENNIAL VEGETABLES

- **Artichoke.** Cut flower buds from plant while tight and firm (artichokes become tough once buds start to open).

- **Asparagus.** Break spears at soil level when they're 7 to 9 inches tall and the tips are still tight (spears toughen when the tips open and begin to fern out). Harvest every other day for best quality. Harvest for 3 weeks after initial emergence during the second season the plant is in the ground, and for 6 to 8 weeks the third season after planting and all following years. Once the harvest window is over, allow plants to fern out and grow to maturity to feed the roots and preserve the longevity of the planting.

- **Cardoon.** Blanch stems for 3 to 4 weeks prior to harvest by wrapping them in burlap, landscape fabric, or other materials. Cut stems with a knife or shears, remove leaves and thorns with a knife or scissor, then use a vegetable peeler to remove the majority of the tough fibers on the exterior of the stem.

- **Jerusalem artichoke, sunchoke.** Dig tubers in late fall, midwinter, or early the following spring. Flavor is best if tubers have experienced cold temperatures.

- **Rhubarb.** Break larger outer stalks from plant and remove most of leaf from stalk with a knife or shears (a tiny bit of leaf left on the stem will keep it crisp during storage). Rhubarb leaves are toxic; make sure to remove them completely before cooking. Pluck about two-thirds of the stalks from a given plant, then wait 5 weeks before next harvest (you can also selectively harvest individual stalks as needed). In the northern parts of North America, expect two to three harvests per year (two in spring, possibly one in fall). Regrowth slows in summer as temperature increases. Don't harvest rhubarb the year of planting; during the second season, harvest lightly; full harvest begins the third season.

Garlic scapes on hardneck garlic (left) provide an early harvest, and their removal helps the plant produce larger heads for the primary harvest. Male squash blossoms (right) can be harvested alongside the pollinated female fruits.

harvesting "hidden" crops

↓ In addition to the obvious harvests that you have been looking forward to all year, your garden, yard, and neighborhood may contain a few hidden surprises for you. To keep working toward your highest-yield garden, look around for extra harvests wherever you can.

EATING THE WHOLE CROP

Many crops, including some that you will be growing in your garden, contain extra sustenance if you become more creative with their preparation. For example, carrot tops are delicious in soups, and beet leaves can be prepared just like kale or chard. Bolted cilantro can be kept in the garden until seed heads form for a harvest of coriander;

broccoli leaves can serve as a replacement for kale or cabbage when you are in a tight spot for cooking greens; squash blossoms are a summertime delicacy; and even grape leaves are edible.

Keep in mind that not all parts of all crops are edible. For example, don't eat tomato leaves or rhubarb leaves. However, as you get to know all of the plants in your space, you'll be amazed at how many hidden foods are already available.

WEEDS

This might already be obvious to you, or it may be a revelation: Many weeds in and around your garden are very edible, and often nutritious and tasty. Chickweed may grow rampant in the garden. Why not harvest it and add it to salads? Many people will pay top dollar for dried nettle leaves, but it can often be harvested all around the neighborhood. Young dandelion leaves are surprisingly tasty when mixed with kale or other cooking greens. If your goal is to maximize the food-production potential of your space, identifying and using free-growing weeds is an effortless way to increase your harvest.

Be careful to harvest weeds only from sites you know to be free of chemical usage. Also, be sure to harvest from sites not used for animal wastes. If your dog uses the lawn for a bathroom, avoid eating edible weeds from this area. You might want to limit harvest of weeds to your own yard or properties you know to be safe. Weeds along roadsides and even in city parks may be sprayed with herbicides.

Common weeds such as the dandelion are nutritious and delicious.

OTHER PLANT PARTS TO EAT

- × **Arugula:** flowers, seedpods
- × **Basil:** flowers
- × **Beans, fava (broad):** shoots, young leaves, flowers
- × **Beets:** leaves
- × **Broccoli:** secondary florets, flowers, flower stems and leaves
- × **Broccoli raab:** flowers
- × **Brussels sprouts:** top leaf cluster, leaves
- × **Cabbage:** outer leaves
- × **Cabbage, Chinese:** outer leaves
- × **Carrots:** greens (tops)
- × **Cauliflower:** leaves
- × **Celeriac:** tops
- × **Chard, Swiss:** stalks

- × **Cilantro:** flowers, seeds
- × **Collards:** flowers
- × **Cucumbers:** flowers
- × **Dill:** fronds, flowers, seeds
- × **Fennel, bulbing:** fronds
- × **Garlic:** scapes (flower stalks)
- × **Kale:** flowers
- × **Kohlrabi:** leaves
- × **Leeks:** flowers
- × **Mustard greens:** flowers
- × **Okra:** flowers
- × **Onions, bulb:** tops, flowers
- × **Pak choi:** flowers, flower stems
- × **Parsley:** flowers
- × **Peas, shelling:** growing tips, flowers, tendrils, shoots

- × **Peas, snap:** growing tips, flowers, tendrils, shoots
- × **Radishes:** flowers, seedpods
- × **Rutabagas:** flowers, seedpods
- × **Scallions:** flowers
- × **Squash, gourds:** flowers
- × **Squash, pumpkins:** flowers, leaves
- × **Squash, summer:** flowers, leaves
- × **Squash, winter:** flowers, leaves
- × **Sweet potatoes:** leaves
- × **Turnips:** flowers
- × **Watermelon:** flowers

COMMON EDIBLE WEEDS

PLANT	LATIN NAME	COMMONLY EATEN PARTS
Bamboo shoots	*Bambusa* species	Shoots
Burdock (and other docks)	*Arctium* species	Roots
Chickweed	*Stellaria media*	Leaves, flowers
Chicory	*Cichorium intybus*	Roots
Clovers	*Trifolium* species	Leaves
Dandelion	*Taraxacum officinale* (and other species)	Leaves, flowers, roots
Daylily	*Hemerocallis* species	Bulbs
Lamb's quarters	*Chenopodium album*	Leaves
Miner's lettuce	*Claytonia perfoliata*	Leaves, flowers
Nettle	*Urtica dioica*	Young leaves
Pigweed	*Amaranthus* species	Leaves, seeds
Plantain	*Plantago major*	Leaves
Purslane	*Portulaca oleracea*	Leaves
Sorrel	*Rumex acetosa*	Leaves
Watercress	*Nasturtium officinale*	Leaves, flowers
Wild asparagus	*Asparagus officinalis*	Shoots
Wild mustard	*Brassica* species	Leaves, flowers
Wild onions	*Allium* species	Leaves, bulbs

EXISTING BUT UNDERUTILIZED EDIBLES

Walking your neighborhood and surrounding locales provides good exercise and the opportunity to identify existing edible crops. Fruit trees can be common in established neighborhoods, and it is possible that much or all of the fruit goes unharvested each year. Making a map of these opportunities, identifying the owner and getting permission to harvest (if applicable), and keeping tabs on the plant through the season to make sure you don't miss peak harvesttime can pay off in spades.

You might find a single apple tree near your house that can provide you with months of fresh apples and a year's supply of applesauce, all for a few hours of harvesting! Pay back the favor and improve your yield by pruning the tree and cleaning up dropped fruit in summer.

All crops should be washed and handled responsibly to prevent contamination.

harvest safely

As a sustainably minded production gardener, you have the opportunity to enjoy some of the safest food in the world. You have control of every step of the growing process—choosing what goes into the soil, what you use to care for the crops, and how to harvest them. Your crops won't ever be transported long distances or moved through multiple warehouses, so chances for outside contamination are slim.

However, just because you're a small producer using environmentally sound techniques doesn't mean you can turn a blind eye to food safety. This is especially true if you're sharing crops with friends, donating them to a food bank, or selling them to a restaurant down the street. You'll need to keep crops free of pathogens.

AVOIDING FECAL CONTAMINATION

The main pathogenic concern as production gardeners is fecal matter, both human and animal. Fecal matter is a well-known carrier of many pathogens, so obviously it's best if you can keep it off of your food. Potential sources of contamination could be your hands, tools, containers, pets,

Wash your hands before harvesting crops to help keep food clean and safe for eating.

wildlife, dirty water, poorly composted manure, or boots/shoes/garden clogs/other fashionable garden footwear. It's worth noting that you should be principally concerned with crops that will be eaten raw (e.g., salad mix or carrots). Cooking or processing foods eliminates the vast majority of common pathogens.

Hands, tools, and containers. This is an easy one. Wash your hands before harvesting. Wash them for a full 30 seconds, and use soap and warm water. While you're at it, wash your harvest knife and harvesting containers, especially if you're harvesting something that will be eaten raw. Consider rinsing them with a sanitizing agent. Diluted bleach works well, or use an organically approved sanitizer.

Pets/wildlife. A big pile of dog poo in the garden does not count as extra fertilizer. If you find feces in your garden, remove it and the soil that it's been in contact with and wash your hands. Bury it far from the garden or put it in the trash. Don't put it in your home compost pile. If this becomes a recurring problem, consider fencing your garden to keep pets and wildlife out.

Dirty irrigation water. If you're irrigating with municipal water, you don't need to be too concerned about it as a source of pathogens. If you're irrigating from a well, we recommend testing it for coliform about once a year. You'll want to do this anyway if it's your drinking water source. If you're irrigating with water from a pond, stream, or collected in a rain barrel from your roof, you'll definitely want to test it, because it has the potential to become contaminated. Consider using drip irrigation to keep pathogens off the leaves of crops that will be eaten raw.

Dirty washing water. You should only use potable water for washing crops. Avoid soaking crops in water because they can draw pathogens into their tissue, especially if the water is more than 10°F (12°C) colder than the produce. A better method is to spray the crops clean on a mesh table.

Poorly composted manure. Manure has the potential to carry pathogenic bacteria, which is why we recommend composting it completely before applying it to the garden. Many home compost piles won't ever reach the desired temperature of 140°F (60°C). If you're concerned that you haven't gotten your pile hot enough, it's still safe to apply it to your garden as long as you wait 4 months before planting crops in that space. This is a good reason to apply your compost in fall and let it age before spring planting.

Footwear. Consider keeping a dedicated pair of shoes for the garden so you don't track anything undesirable into your growing space.

harvesting for maximum freshness & quality

↓ Your produce is undergoing cellular respiration while it's attached to the plant, and it continues to do so after you pick it. To put that another way, your food is still alive after you harvest it. This is a cool concept, to be sure, but while you're thinking about how cool it is, it's already causing your produce to break down and lose quality.

It's important to slow down the process by cooling your produce to an appropriate temperature as quickly as possible. For some crops (winter squash, onions, garlic), you'll need to cure them at a higher temperature for a short period to reduce moisture levels, then move them into colder storage for long-term keeping (more on this later).

Some vegetable crops have much higher cellular respiration rates than others, so they deteriorate more quickly after harvest. These are the most important crops to cool quickly and keep at an appropriate temperature to preserve quality. These are also the crops that are best eaten as soon as possible after harvest.

You may have already learned this fact through the Sweet Corn Experience (distantly related to the Jimi Hendrix Experience). Sweet corn purchased from a farmers' market or a roadside stand that was picked that morning is always better than corn that's been sitting in a grocery store for a few days. Sweet corn is a high-respiration crop, which is why it tastes best if eaten the day that it was harvested. Otherwise it

Harvesting lettuce when temperatures are cool ensures the greens will store as long as possible.

Allow crops like dry beans to cure as much as possible in the field before harvest.

should be very quickly cooled to 33°F (0.5°C) and held there. High-respiration crops can be very satisfying to grow at home because you get to eat them at their peak.

DAILY HARVEST TIMING

One of the easiest ways to ensure postharvest quality is to pick your crops at an appropriate time of day. If a crop needs to be cooled to preserve its postharvest quality, it makes sense to harvest it when the air temperature is cooler. You may have already learned this by trial and error. If you cut a head of lettuce or pick a bunch of kale at noon when it's 80°F (27°C) outside and leave them in the sun for even a few minutes, they'll quickly wilt. However, if you pick them at seven in the morning when the temperature is 60°F (16°C), they'll hold their quality much longer.

This happens because plant metabolic activity is higher when air temperatures are higher. When you pick lettuce midday, its cells are cranking away to help the plant grow, and they're still cranking away after you remove its water supply by cutting it from the roots. Conversely, if you pick lettuce early in the morning, the cells are still relaxing in bed and drinking coffee, so the leaves don't break down as quickly after cutting.

Additionally, many crops store best at high humidity levels because they stay hydrated and crisp. If you pick these crops when there's still dew on them, it's easier to maintain an appropriate humidity level after harvesting. Some large commercial lettuce operations actually do their harvesting at night because it saves them a lot of time and energy that would otherwise need to be spent on cooling and hydrating the lettuce.

In contrast to lettuce, some crops keep best when harvested after the dew has burned off, like tomatoes and cucumbers. These crops can be more prone to rot after harvest if their outer surface is wet.

A good strategy is to pick crops that like to be cool and wet early in the morning, and to pick crops that like to be dry in the middle of the day or evening when the dew has burned off.

Remember, these are suggestions, not rules. You'll most likely do your harvesting whenever you have time, but not to worry: Crops can easily be cooled, hydrated, or dried off after harvest.

HARVESTING CROPS THAT LIKE IT COOL & WET

For small quantities, all you need to do is harvest your crops when it's cool outside and there's still dew on the leaves, place them in a plastic bag, and set them in the fridge. The bag keeps the humidity high, and the fridge quickly cools things down. Be careful not to jam too much produce into a small drawer or compartment; air circulation around the bags is important to cool the crops. If you need to keep the produce outside for a while, bag it to maintain humidity and be sure to keep it in the shade. You might give each bag a spritz of water if the crop is completely dry at harvesttime.

If the crops are dirty and/or were harvested during the heat of the day, place them in a colander or on a drain table (see Washing & Cooling below) and give them a good rinsing. This will clean them up and cool them down. For small batches of mixed lettuces and salad greens like arugula, mizuna, or baby spinach, a salad spinner works well for drying before packing. For larger quantities, fill a mesh laundry bag and spin it around by hand. Be sure to wash and dry the mesh bag between uses.

Crops to harvest cool and wet. These include salad greens, cooking greens, leeks, scallions, sweet corn, all brassicas (broccoli, cabbage, cauliflower, kale, pak choi, mustard greens), and all root crops with tops (beets, carrots, radishes, rutabagas, turnips).

HARVESTING CROPS THAT LIKE IT DRY

These crops are pretty easy—all you need to do is wait until the dew has dried, then harvest away. If the crops need to be cleaned or are still a little damp with dew, most of them can be wiped off with a dry rag, or wear soft cotton gloves while you're harvesting and wipe them off as you pick. Most crops that like to be dry also don't mind being a little warmer at harvesttime. If you'll be eating them soon, don't worry about cooling solanums and cucurbits; just pick them and store them at room temperature until they're ready

to use. Beans do appreciate a cooling rinse if it's above 70°F (21°C). Onions and garlic, and sometimes potatoes, need to be cured before storage (see page 260); most growers will wait until after curing to clean them.

Crops to harvest dry. These include solanums (potatoes, eggplant, peppers, and tomatoes), cucurbits (cucumbers, melons, and squash), beans, berries, storage onions, and garlic.

WASHING & COOLING

The primary cleaning equipment of the home garden is the garden hose. The hose also serves as a great cooling tool: Using cold water to cool produce is called hydrocooling. Make sure you use a drinking-water approved hose; most common garden hoses contain bactericides and unacceptable levels of lead.

For most home production growers, the primary cooling and storage equipment is the kitchen refrigerator. If you're growing large amounts of produce or will be sharing or distributing your harvest, you might consider buying an auxiliary fridge, or even setting up a small walk-in cooler to hold your harvests.

If you need to wash/cool large quantities of produce, consider building a drain table. Vinyl-coated welded wire fencing lets dirty water drain through and is easy to clean. If you want to get fancy, you can purchase a manufactured plastic or stainless steel drain table.

TUB WASHING

We generally don't recommend soaking produce in a tub to clean it. Immersing into water can cause produce to internalize pathogens, and it allows pathogens to spread from one piece of produce to many.

That said, tub washing is a useful technique that is often used on professional organic farms to efficiently wash baby lettuce and salad greens, and it can be safely done using two or three wash

A wash pack station can be simple, but should be clean and organized.

tubs. Professional growers often add an organically approved sanitizer to keep pathogens from multiplying and spreading in the water.

On a small scale, almost all garden produce can be effectively cleaned and cooled using a colander or drain table. If you need to wash or cool a larger amount of baby greens, you may want to use wash tubs.

To prepare for tub washing, select two or three plastic or stainless steel tubs appropriately sized for your harvest. If using plastic, select #2 plastic, which is inert and won't leach or react with the sanitizer. Using three separate tubs is the gold standard (triple washed), but using two tubs is also acceptable. Make sure they're clean before use. Along with the tubs, you'll need an organically approved sanitizer such as SaniDate or Tsunami 100, if desired, and a colander.

Wash your hands before getting started, then follow these steps:

+ Fill two or three tubs with potable water.

+ Add sanitizer as directed on the package, if desired. Some growers use sanitizer in all three tubs; some use it only in the third and

final rinse; some use it in the second; some use it in the first; and some don't use it at all. We prefer a triple-rinse system and use sanitizer in the first two tubs. We feel the risk of pathogen transfer is greatest in these tubs, and we like to completely rinse the sanitizer off in the final tub.

+ Place greens into the first tub. Stir gently, then use the colander to move them into the second tub. Stir again and move to the third tub, if using.

+ Stir gently, then use a colander to drain the greens and place them into your desired spinning apparatus to dry.

HARVESTING TOOLS

In addition to the tools you'll use during the growing season, you'll probably want to keep some specialized tools and equipment on hand for harvesting. Here's what we recommend.

CUTTING IMPLEMENTS

Scissors are a great harvest tool for salad greens, soft herbs, and any other thin plant matter.

Knives are great for salad greens, head lettuce, and cutting through thicker plant stems (broccoli, cabbage, cauliflower). Pruning shears are excellent for cutting through tough stems on summer and winter squash, cucumbers, tomatoes, and peppers. We recommend keeping a dedicated pair of each of these tools for harvesting to avoid contamination of produce, especially for crops that will be eaten raw.

HARVESTING CONTAINERS

You'll need a container to carry your produce back to the house. We love traditional baskets for harvesting a wide variety of produce. A metal colander is great for washing and cooling crops, and a large plastic tub is useful for carrying heavy crops like squash in from the garden.

SCALE

This is not essential, but is useful for keeping accurate harvest records or if you're selling any of your crops. A simple kitchen scale works well, or you can purchase a digital produce scale if you want more accurate measurements.

HOSE FOR WASHING PRODUCE

Most common garden hoses contain surprisingly high levels of lead, so we recommend finding a lead-free hose to use for produce washing. Nowadays, lead-free hoses are relatively easy to come by, but it's important to check, rather than assume. Manufacturers are eager for you to know about their new lead-free hoses, so look around and find one that you feel good about.

SOAP/HAND SANITIZER

You should always wash/clean your hands before handling produce. Keep a small bottle of biodegradable soap or hand sanitizer with your harvest tools. These can also be used to clean your cutting implements, containers, and the top of your scale.

PRODUCTION-SCALE TOOLS

If you are scaling up your operation, there are a number of specialty tools available that are used by small farmers to make their harvesting and produce-washing systems more efficient. These include commercial-size salad spinners, handheld mechanical greens harvesters, barrel washers to quickly clean large quantities of root crops, and large harvesting tubs for picking bulk amounts of produce. Most of these tools will probably exceed your needs, but if you find that the harvesting and processing of a particular crop is taking lots of your time, they can be worth looking into. Many growers also choose to design and build their own tools to meet their specific needs.

Every grower has their preferred harvest tool. Shown here, lettuce field knives and serrated greens knives.

Beets stored in a plastic bag in the refrigerator can keep for 4 months or more.

storing the harvest

↓ Crop storage is an important skill that allows you to enjoy garden produce over long periods of time. With appropriate storage techniques, you can still be eating your produce weeks or months after you harvest it. Some crops can be stored whole and cooked later. Whole crops usually need some type of temperature-controlled space such as a refrigerator, walk-in cooler, or root cellar. A few crops can even be successfully stored in the garden buried under the soil during winter months.

Certain crops can be stored more efficiently, for longer periods, or in a tastier form by processing them. When we talk about processing, we're

not talking about adding MSG, hydrogenated oils, or artificial preservatives—we're talking about the time-honored techniques of canning, freezing, drying, and pickling.

When large-scale growers are storing crops for extended periods, they might have three or four walk-in coolers that are set to precise temperature and humidity levels. This allows them to store and sell crops over the longest period possible. Most gardeners don't have the space or equipment for this, but you can actually do a pretty good job using what you already have at home.

STORAGE LOCATIONS

The specific crops you are storing, the types of space you have available, and your end use will dictate which storage option is most appropriate for a particular harvest.

IN THE REFRIGERATOR

The home refrigerator is the best place to store crops that like it cold (33 to 39°F/0.5 to 4°C) and humid (90 to 100% relative humidity). Placing crops in plastic bags helps maintain high levels of humidity. If crops are very dry when bagging, you can spray a quick spritz of water in the bag. You don't want standing water in the bag, just enough to moisten the crop.

As you scale your garden up, consider a second refrigerator for long-term produce storage. We encourage you NOT to buy the oldest, cheapest fridge you can find; most old refrigerators are very inefficient and have high monetary and environmental operating costs.

ON THE COUNTER

For short-term storage of crops that like warmer temperatures and low humidity, simply placing them in a bowl at room temperature works well. Try this with tomatoes, cured onions and garlic, cucumbers, and summer and winter squash.

IN THE GROUND

A few crops, namely carrots and potatoes, can be stored over the winter outside in the garden. The key is to keep the ground and the roots from freezing. The best way to do this is to cover the planting with a thick layer of mulch (8 inches or more) before air temperatures drop below freezing.

Even with the mulch, this technique may not work in areas with extremely cold winter temperatures. In warmer climates where the soil doesn't freeze during winter, this technique works but crops may be more susceptible to pest problems as they sit in garden storage. For example,

MAKING YOUR OWN COOLER

Many small farmers have begun using a new piece of technology called the CoolBot. This device allows you to convert a normal air conditioner into a refrigeration unit. To use a CoolBot, you'll need a working air conditioner and a space that can be insulated and converted into a walk-in-style cooler. Possible opportunities are basement rooms, sheds, portions of a garage, or used steel shipping containers. If you're harvesting and storing large amounts of produce and want to consider this tool, see Resources (page 292) for more information.

in the Pacific Northwest, potatoes stored in the ground for too long become a tasty treat for marauding wireworms.

IN A ROOT CELLAR

Many people consider the root cellar to be an antiquated piece of farm history, but in truth it is an ingenious, time-saving, low-energy storage system. Traditionally, a root cellar was simply a framed-in hole dug into the ground or the side of a hill. Geothermal energy and insulation from the earth moderated the temperature in the cellar, keeping it cooler in summer and (ideally) above freezing in winter. If you have the space, time, and skills to construct one, you can give it a try. Be sure to install vents to maintain air circulation and allow cooler outdoor air to enter.

In your house, the simplest "root cellar" might be a corner of a basement or a spot in a garage or shed that stays cooler than the rest of your house, but doesn't get below freezing. A basement or heated garage works well for crops that like warmer temperatures and low humidity, such as cured potatoes, sweet potatoes, onions, garlic, and winter squash. These low-humidity crops keep best in well-ventilated containers like

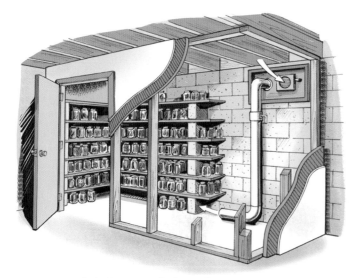

A dry basement is the perfect spot to create an inexpensive food cellar. Providing a way for warm air to escape and cool air to enter will help root vegetables and other unprocessed foods last longer.

wooden crates or cardboard boxes with vent holes in the sides. To discourage crops from sprouting, it is important to keep the space dark. This is especially true for potatoes, which can develop a toxic green coloration if exposed to light. If this occurs, peel them before use to remove any trace of the green coloration.

If it's cold in your garage or shed, you can use it as a space to store root crops that like lower temperatures and higher humidity like carrots, beets, radishes, and turnips. These crops ideally prefer temperatures 30 to 40°F (–1 to 4°C), but will keep for a while even in temperatures a little above 50°F (10°C). Store them in plastic bags or bins to reduce water loss.

Buried containers. Another take on a root cellar is simply to bury a container underground for storing root vegetables. Try burying a food-grade plastic bucket or drum with a lid so that the rim protrudes a few inches above ground. Fill it with produce packed between layers of sawdust, clean straw, or peat moss. Secure the lid, and cover it with 8 or more inches of soil or straw mulch. Root crops such as beets, carrots, or turnips are probably best for this type of storage. This method is not perfect because you have no control of air circulation or temperature, but it works.

CURING FOR LONG STORAGE

Certain crops can be cured before eating and storing to improve flavor and increase storage life. Curing is the process of holding a crop at a relatively high temperature for a short period after harvest. Because the curing process varies by the crop, we'll discuss crops individually.

ONIONS

If you harvest onions while the tops are still green, you can use both the bulbs and tubular leaves for raw eating or cooking. However, to store onions for more than a few weeks it's necessary to cure them. To do so, stop irrigating when bulbs are mature and the tops are starting to die back. Wait until the tops are beginning to brown and fall over, then pull the bulbs from the ground. If the weather is forecasted to be warm and sunny, you can leave the pulled onions right in your beds in the sun for several days to begin the curing process. If it's extremely hot or rain is in the forecast, skip this step.

Store the onions, with tops and roots still attached, in a dry, well-ventilated location until the tops are fully brown and dried. This should take about 2 to 3 weeks. At this point, you can trim the roots and tops to within 1 inch of the bulb, and move them into long-term storage. A greenhouse with the doors open is a great place for curing. Ideal curing temperature is 75 to 80°F (24 to 27°C), but lower temperatures are fine if

Lay out onions in a warm area with good air circulation to finish curing before storage.

you don't have heated space or the weather isn't cooperating. If you have room, you can spread the onions on racks or a mesh-top table in a single layer for maximum airflow; bundling the onions by their tops and hanging under cover also works well.

GARLIC

Fresh-picked garlic has a unique, mild flavor before curing that you never get to experience with store-bought garlic, so we encourage you to eat some fresh. As with onions, garlic must be cured for long-term storage. Withhold water when bulbs are approaching maturity. Harvest garlic when about half the leaves are brown and half are still green. Don't wait until all the leaves are dead or storage life and flavor will be compromised. After harvesting, follow the steps for curing onions, but omit the optional few days of curing in the field, as direct sun exposure after harvest can break down flavor compounds. Softneck garlic can be braided together and hung when the tops are fully dry for an attractive storage option.

WINTER SQUASH & STORAGE PUMPKINS

Winter squash does not need to be cured before eating, but it will keep best after curing for 7 to 10 days at 70 to 80°F (21 to 27°C). For best flavor, wait to harvest until the stem is dry and the fruits have a yellowish/brownish mark where they've been in contact with the ground. If you need to harvest your squash early to protect it from rain or adverse weather in fall, curing it can greatly improve flavor. Wipe down squash fruits with a very dilute bleach solution (1 tablespoon/gallon of water) to eliminate fungal spores and extend storage life.

POTATOES

"New potatoes" are harvested before the top growth has started to brown and are eaten

A NOTE ABOUT ETHYLENE

Ethylene is a gas that serves as a plant growth regulator. It's naturally produced by plants, and helps certain crops continue to ripen after harvest. However, it causes problems with vegetables in storage because it causes many crops to rot prematurely. For long-term storage, notorious ethylene producers such as apples, bananas, cantaloupe, pears, and tomatoes should be stored separately from other vegetable crops. Conversely, you can use ethylene producers to your advantage—if you want to hasten the ripening process for greenish tomatoes harvested at the end of the growing season, try storing them with a few apples.

uncured. These early potatoes have amazing flavor and texture. However, for long-term storage, potatoes should be cured to slightly toughen the skin and heal any wounds from harvesting. Dig up potatoes when the tops have died back. Before washing, hold them in a dark, well-ventilated space at 55 to 60°F (10 to 15°C) for 1 to 2 weeks. For longest storage life, don't wash until immediately before use. For moderate-term storage, you can wash after curing and then air-dry. Cleaning prior to storage keeps things a little tidier if your storage area is in a kitchen cupboard, garage, or basement. You can also store potatoes in the ground (see page 259).

SWEET POTATOES

For long-term storage, dig sweet potatoes after the foliage has started to yellow and die back. You can leave them in the field exposed to the sun for a few hours to begin the curing process, if conditions are favorable. Before washing, cure at 70 to 90°F (21 to 32°C) in a well-ventilated space for about 2 weeks.

CROP STORAGE

ANNUAL VEGETABLES & HERBS	IDEAL STORAGE TEMP. (°F)	SUGGESTED STORAGE % RELATIVE HUMIDITY	PREFERRED STORAGE CONTAINER	BEST HOME STORAGE PRACTICE
Arugula	33	95	Plastic bag/liner	Bagged in refrigerator
Basil	50	90	Paper bag/plastic bag with vents	Paper bag at room temperature
Beans, edible soy (edamame)	33	95	Plastic bag/liner	Bagged in refrigerator
Beans, fava (broad)	41–45	95	Plastic bag/liner	Bagged in refrigerator
Beans, lima	41–45	95	Plastic bag/liner	Bagged in refrigerator
Beans, shell (fresh)	41–45	95	Plastic bag/liner	Bagged in refrigerator
Beans, shell (dried)	70	65	Paper bag/glass jar	At room temperature in dark cupboard
Beans, snap	41–45	95	Plastic bag/liner	Bagged in refrigerator
Beets	33	95	Plastic bag/liner	Bagged in refrigerator
Broccoli	33	95	Plastic bag/liner	Bagged in refrigerator
Broccoli raab	33	95	Plastic bag/liner	Bagged in refrigerator
Brussels sprouts	33	95	Plastic bag/liner	Bagged in refrigerator
Cabbage	33	95	Plastic bag/liner	Bagged in refrigerator
Cabbage, Chinese	33	95	Plastic bag/liner	Bagged in refrigerator
Carrots	33	95	Plastic bag/liner	Bagged in refrigerator
Cauliflower	33	95	Plastic bag/liner	Bagged in refrigerator
Celeriac	33	95	Plastic bag/liner	Bagged in refrigerator
Celery	33	95	Plastic bag/liner	Bagged in refrigerator
Chard, Swiss	33	95	Plastic bag/liner	Bagged in refrigerator
Cilantro	33	95	Plastic bag/liner	Bagged in refrigerator
Collards	33	95	Plastic bag/liner	Bagged in refrigerator
Corn, sweet	33	95	Plastic bag/liner	Bagged in refrigerator

ANNUAL VEGETABLES & HERBS	IDEAL STORAGE TEMP. (°F)	SUGGESTED STORAGE % RELATIVE HUMIDITY	PREFERRED STORAGE CONTAINER	BEST HOME STORAGE PRACTICE
Cucumbers	50–55	95	Paper bag/plastic bag with vents/wrap in paper towel/on the counter	At room temperature, with or without paper bag
Dill	33	95	Plastic bag/liner	Bagged in refrigerator
Eggplant	50–55	90	Paper bag/plastic bag with vents/wrap in paper towel/on the counter	At room temperature, with or without paper bag
Endive	33	95	Plastic bag/liner	Bagged in refrigerator
Fennel, bulbing	33	95	Plastic bag/liner	Bagged in refrigerator
Garlic	33	60–70	Mesh bag/well-ventilated box/on the counter	At room temperature in well-ventilated mesh bag
Kale	33	95	Plastic bag/liner	Bagged in refrigerator
Kohlrabi	33	95	Plastic bag/liner	Bagged in refrigerator
Leeks	33	95	Plastic bag/liner	Bagged in refrigerator
Lettuce, baby mix	33	95	Plastic bag/liner	Bagged in refrigerator
Lettuce, head	33	95	Plastic bag/liner	Bagged in refrigerator
Mâche	33	95	Plastic bag/liner	Bagged in refrigerator
Melon, cantaloupe, honeydew	36–41	95	Well-ventilated cardboard box or none	Unbagged in refrigerator or on the counter
Mustard greens	33	95	Plastic bag/liner	Bagged in refrigerator
Okra	45–50	95	Paper bag/plastic bag with vents/wrap in paper towel/on the counter	At room temperature
Onions, bulb	33	60–70	Mesh bag/well-ventilated cardboard box/on the counter	At room temperature in well-ventilated location

ANNUAL VEGETABLES & HERBS	IDEAL STORAGE TEMP. (°F)	SUGGESTED STORAGE % RELATIVE HUMIDITY	PREFERRED STORAGE CONTAINER	BEST HOME STORAGE PRACTICE
Pak choi	33	95	Plastic bag/liner	Bagged in refrigerator
Parsley	33	95	Plastic bag/liner	Bagged in refrigerator
Parsnips	33	95	Plastic bag/liner	Bagged in refrigerator
Peanut	34–41	55–70	Paper bag	Room temperature for a few months, refrigerated for up to a year
Peas, shelling	33	95	Plastic bag/liner	Bagged in refrigerator
Peas, snap	33	95	Plastic bag/liner	Bagged in refrigerator
Peppers, hot	41–50	85–95	Paper bag/well-ventilated cardboard box/on the counter	At room temperature
Peppers, sweet	45–50	90–95	Paper bag/well-ventilated cardboard box	At room temperature
Potatoes	40–50	90–95	Paper bag/well-ventilated cardboard box	At room temperature in the dark
Radicchio	33	95	Plastic bag/liner	Bagged in refrigerator
Radishes	33	95	Plastic bag/liner	Bagged in refrigerator
Rutabagas	33	95	Plastic bag/liner	Bagged in refrigerator
Scallions	33	95	Plastic bag/liner	Bagged in refrigerator
Spinach	33	95	Plastic bag/liner	Bagged in refrigerator
Squash, gourds	50–55	50–75	Paper bag/well-ventilated cardboard box/on the counter	At room temperature
Squash, pumpkins	54–59	50–75	Paper bag/well-ventilated cardboard box/on the counter	At room temperature

ANNUAL VEGETABLES & HERBS	IDEAL STORAGE TEMP. (°F)	SUGGESTED STORAGE % RELATIVE HUMIDITY	PREFERRED STORAGE CONTAINER	BEST HOME STORAGE PRACTICE
Squash, summer	41–50	95	Paper bag/well-ventilated cardboard box/on the counter	At room temperature
Squash, winter	50–55	50–75	Paper bag/well-ventilated cardboard box/on the counter	At room temperature
Sweet potatoes	55–59	85–95	Paper bag/well-ventilated cardboard box/on the counter	At room temperature
Tomatillos	45–55	85–95	Paper bag/well-ventilated cardboard box/on the counter	At room temperature
Tomatoes	45–55	85–95	Paper bag/well-ventilated cardboard box/on the counter	At room temperature
Turnips	33	95	Plastic bag/liner	Bagged in refrigerator
Watermelon	50–59	90	Well-ventilated cardboard box	At room temperature

PERENNIAL VEGETABLES	IDEAL STORAGE TEMP. (°F)	SUGGESTED STORAGE % RELATIVE HUMIDITY	PREFERRED STORAGE CONTAINER	BEST HOME STORAGE PRACTICE
Artichoke	32	95	Plastic bag/liner	Bagged in refrigerator
Asparagus	35	95	Plastic bag/liner	Bagged in refrigerator
Cardoon	35	95	Plastic bag/liner	Bagged in refrigerator
Jerusalem artichoke, sunchoke	33	95	Plastic bag/liner	Bagged in refrigerator
Rhubarb	33	95	Plastic bag/liner	Bagged in refrigerator

GARDEN-PLANNING WORKSHEETS & CHARTS

We developed these worksheets so you have all the information you need to build your own planting calendar and garden plan. For those of you that prefer a digital format, you can download all of the worksheets on our website at www.seattleurbanfarmco.com/worksheets.

CROP AMOUNT WORKSHEET

To determine the number of servings you want to produce per week, consider how many people you will be growing for and how many servings they will eat each week. You can also account for extra amounts for freezing and/or canning in this column. Please note that the weights and volumes in the Average Serving Size column are for raw produce (before cooking). For more on planning for crop storage, see page 30. To determine the number for the Total Weight of Crop column, multiply the number of ounces in the Average Serving Size column with the number in the Number of Servings column, and then divide by 16 (because there are 16 ounces in a pound).

ANNUAL VEGETABLES	AVERAGE SERVING SIZE	NUMBER OF SERVINGS YOU WANT TO PRODUCE EACH WEEK	TOTAL WEIGHT OF CROP NEEDED PER WEEK IN POUNDS
Arugula	4 ounces		
Basil	8 leaves or 1 ounce		
Beans, edible soy (edamame)	1 cup or 4 ounces, unshelled		
Beans, fava (broad)	8 ounces unshelled beans		
Beans, lima	8 ounces fresh, in pods		
Beans, shell (fresh)	20 pods		
Beans, shell (dried)	¼ cup		
Beans, snap	1 cup or 3.8 ounces		
Beets	½ cup or 3 ounces		
Broccoli	¾ cup or 2.4 ounces		
Broccoli raab	¾ cup or 2.4 ounces		
Brussels sprouts	6 sprouts or about 4 ounces		
Cabbage	1 cup or 3 ounces		
Cabbage, Chinese	1 cup or 3 ounces		
Carrots	1 medium-size root or 2.2 ounces		
Cauliflower	½ cup or 2 ounces		
Celeriac	½ cup or 2.5 ounces		
Celery	½ cup or 2 ounces		
Chard, Swiss	3 cups raw, chopped or 8 ounces		

ANNUAL VEGETABLES	AVERAGE SERVING SIZE	NUMBER OF SERVINGS YOU WANT TO PRODUCE EACH WEEK	TOTAL WEIGHT OF CROP NEEDED PER WEEK IN POUNDS
Cilantro	Used as spice, about 0.5 ounce		
Collards	3 cups raw, chopped or 8 ounces		
Corn, sweet	1 ear or about 2.2 ounces kernels		
Cucumbers	½ cup or 8 ounces		
Dill	Used as spice, about 0.5 ounce		
Eggplant	1 cup or 3 ounces		
Endive	1 cup or 1.7 ounces		
Fennel, bulbing	1 cup or 3 ounces		
Garlic	1 clove or 0.2 ounce		
Kale	3 cups raw, chopped, or 8 ounces		
Kohlrabi	½ bulb or about 4 ounces		
Leeks	1 cup or 8 ounces		
Lettuce, baby mix	4 ounces		
Lettuce, head	2 cups or 2–4 ounces		
Mâche	4 ounces		
Melon, cantaloupe, honeydew	½ cup or 3 ounces		
Mustard greens (full size for braising)	3 cups raw, chopped, or 8 ounces		
Okra	10 pods or 3.5 ounces		
Onions, bulb	½ cup or 3 ounces		
Pak choi	2 cups or 5 ounces		
Parsley	Used as spice, about 0.5 ounce		
Parsnips	4 ounces		
Peanut	1.5–2 ounces		
Peas, shelling	8 ounces unshelled or ½ cup shelled		

ANNUAL VEGETABLES	AVERAGE SERVING SIZE	NUMBER OF SERVINGS YOU WANT TO PRODUCE EACH WEEK	TOTAL WEIGHT OF CROP NEEDED PER WEEK IN POUNDS
Peas, snap	1 cup or 2.2 ounces		
Peppers, hot	Variable, 0.1–1 ounce		
Peppers, sweet	½ cup or 2.6 ounces		
Potatoes	3–4 ounces		
Radicchio	1 cup or 1.4 ounces		
Radishes	½ cup sliced or 2 ounces		
Rutabagas	6–8 ounces		
Scallions	½ cup or 1.7 ounces		
Spinach	1 cup or 1.1 ounces		
Squash, pumpkins	8 ounces		
Squash, summer	1 cup or 4 ounces		
Squash, winter	8 ounces		
Sweet potatoes	½ cup or 2.5 ounces		
Tomatillos	2–4 ounces		
Tomatoes	1 medium-size fruit or 4 ounces		
Turnips	6–8 ounces		
Watermelon	½ cup or 2.7 ounces		

PERENNIAL VEGETABLES	AVERAGE SERVING SIZE	NUMBER OF SERVINGS YOU WANT TO PRODUCE EACH WEEK	TOTAL WEIGHT OF CROP NEEDED PER WEEK IN POUNDS
Artichoke	1 artichoke or 4.5 ounces		
Asparagus	1 cup or 4.5 ounces		
Cardoon	2 basal leaves or 4 ounces		
Jerusalem artichoke, sunchoke	3–4 ounces		
Rhubarb	1 stalk or 2.5 ounces		

YIELDS OF ANNUAL VEGETABLES & HERBS

Yields can vary greatly, depending on your climate, soil fertility, which cultivars you select, and many other factors. We strongly suggest you keep your own records and adjust your plantings accordingly in future seasons.

Yields are approximate and calculated per row foot. If you are growing using the square-foot method, you can still easily determine how many row feet of each crop you have. For a given crop, take the length of one side of the square in feet and multiply it times the total number of plants you will be setting out. For example, if you will be growing 8 kohlrabi plants in 6" squares, 0.5' × 8 plants = 4 row feet of kohlrabi.

CROP	YIELD PER ROW FOOT	NOTES
Arugula	0.25 pound	Per cutting; expect 1–2 cuttings a week apart.
Basil	0.1 pound	Per week; expect 1–2 harvests per week over the course of the season.
Beans, edible soy (edamame)	0.2 pound	
Beans, fava (broad)	Varies	Expect to harvest 15–60 total pods over the harvest period (closer to 15 for large-seeded varieties, closer to 60 for small-seeded varieties) per plant.
Beans, lima	0.1–0.2 pound	
Beans, shell (fresh or dried)	0.1–0.2 pound	
Beans, snap (bush)	0.25 pound	Per picking; expect 1–4 pickings about 1 week apart.
Beans, snap (pole)	0.25 pound	Per picking; expect 6–12 pickings about 0.5–1 week apart.
Beets	0.5 pound or 2–3 roots	Not including greens.
Broccoli	0.4 pound or 1 head plus side shoots	From cutting of main head; plant will continue to produce smaller quantities of side shoots for 1–4 weeks after main head is cut.
Broccoli raab	0.3–0.5 pound	
Brussels sprouts	0.6 pound	You will probably harvest this amount over the course of 2–3 pickings unless you cut the entire stalk at once.
Cabbage	1.5 pounds or 1 head	
Cabbage, Chinese	1.5 pounds or 1 head	

CROP	YIELD PER ROW FOOT	NOTES
Carrots	1 pound or 4–6 roots	
Cauliflower	0.75 pound or 1 head	
Celeriac	0.75 pound or 1–2 roots	
Celery	0.75 pound or 1 head	
Chard, Swiss	0.3 pound or ⅓ of a bunch per week	
Cilantro	0.5 pound	Per cutting; expect 1–3 cuttings about a week apart.
Collards	0.3 pound or ⅓ of a bunch per week	Per picking; expect about 1 picking per week for 2–5 months.
Corn, sweet	0.5 pound or 1–2 ears	
Cucumbers	0.75 pound or 1–2 fruits per week	Per picking; expect about 1–2 pickings per week for 1–2 months.
Dill	0.5 pound	Per cutting; expect about 1 cutting per week for 2–3 weeks.
Eggplant	0.75 pound or 1–2 fruits per week	Per picking; expect about 1 picking per week for 1–3 months.
Endive	0.75 pound or 1 head	
Fennel, bulbing	0.75 pound or 1 bulb	
Garlic	0.1 pound or 1–2 bulbs	
Kale	0.3 pound or ⅓ of a bunch per week	Per picking; expect about 1 picking per week for 2–5 months.
Kohlrabi	0.4 pound or 2–3 bulbs	
Leeks	1 pound or 2–3 leeks	
Lettuce, baby mix	0.15 pound	Per cutting; expect 1–4 cuttings about a week apart depending on conditions.
Lettuce, head	0.75 pound or 1 head	
Mâche	0.1 pound	
Melon, cantaloupe, honeydew	1.5 pounds	Expect 5–10 fruits total per plant with 1–3 ready to pick each week at 1 plant per 3 feet of row.
Mustard greens	0.7 pound	At large braising size; for baby salad size, see Lettuce, baby mix.
Okra	0.1–0.2 pound	Per picking; expect 2–3 pickings per week, yield is lower at first picking and increases later in the season.

CROP	YIELD PER ROW FOOT	NOTES
Onions, bulb	0.8 pound or 2 onions	
Pak choi	1.3 pounds	For full-size types; yield is less for baby size.
Parsley	0.1–0.2 pound	Per picking; expect 1 picking per week, yield increases later in season when plant is larger.
Parsnips	0.75 pound or 4–6 roots	
Peanut	0.3 pound, green/ 0.1 pound, dry	
Peas, shelling	0.25 pound	Total yield including hulls; expect to get this over the course of 2–3 pickings (1 pound of peas with hulls yields about 1.25 cups shelled peas).
Peas, snap	0.1–0.2 pound	Per picking; pick 2–3 times per week for about a month.
Peppers, hot	0.5–1 pound	Per picking; harvestable season varies by variety.
Peppers, sweet	0.5–1 pound	Per picking; harvestable season varies by variety.
Potatoes	3 pounds	
Radicchio	0.75 pound or 1 head	
Radishes	0.5 pound or 1 bunch	
Rutabagas	1.5 pounds or 2–3 roots	
Scallions	0.3 pound	
Spinach	0.2 pound baby, 0.4 pound full size	Per cutting/picking; expect 1 harvest per week over 2–4 weeks.
Squash, pumpkins	1–3 pounds	Expect 1–5 fruits per plant, depending on the variety.
Squash, summer	1 pound	Per picking; pick 1–2 times per week for 1–2 months.
Squash, winter	1 pound	Expect 3–10 fruits per plant, depending on the variety.
Sweet potatoes	2 pounds	
Tomatillos	0.7 pound	Per picking; pick 1–2 times per week over 1–2 months.
Tomatoes	1 pound	Per picking; pick 1–2 times per week over 1–2 months.
Turnips	0.3 pound or 2–3 bulbs	
Watermelon	2 pounds	Expect 1–2 fruits per plant.

PLANTING CALENDAR WORKSHEET

Y ou can download and print blank calendar templates and a spreadsheet with built-in formulas at Seattle Urban Farm Company's website (see Resources, page 292).

Rows 1 and 2: Fill out the first two rows with the crops you've selected from the Crop Amount Worksheet (page 267) and the varieties you think will be best suited to your climate or taste preferences.

Row 3: Use the number of row feet you calculated on the yield chart (page 270) to fill out this row. For transplants, convert row feet to the number of plants: Consult the spacing column on the Garden Planning Chart on page 284. Divide your desired number of row feet by the plant spacing in feet to get the number of plants to set out. For example, you calculate you need 20 row feet of broccoli. The spacing for broccoli is 1 foot. Twenty row feet ÷ 1-foot spacing = 20 plants. Here's a quick reference for spacing plants less than a foot apart: 8 inches = 0.67 foot, 6 inches = 0.5 foot, 4 inches = 0.33 foot, 3 inches = 0.25 foot, 2 inches = 0.17 foot.

Row 4: Use your average first and last frost dates and the Planting Dates Worksheet (page 274) to determine your planting date. Consult the Garden Planning Chart on page 284 to help decide how many plantings to make and how many weeks apart to space the plantings.

Row 5: Specify in this row whether you'll be transplanting or direct seeding the crop.

Rows 6 and 7: These rows apply only if you'll be growing your own transplants. In Row 6: Note the date to start the seeds. Consult the Planting Dates Worksheet on page 274. Subtract the number of weeks listed for your crop from the planting date you specified in this chart. This is the date you should seed the crop in your nursery. In Row 7: Add 20% to the number of plants you specified in Row 3. This is the number of seeds you should start.

Row 8: Indicate when you actually planted the crop.

Row 9: Indicate how many row feet of the crop you actually planted.

Rows 10 and 11: Note when harvest occurred and how much was harvested.

1	Crop	
2	Variety	
3	Quantity	
4	Expected planting date	
5	Transplant or direct seed?	
6	Seed-starting date	
7	Number of transplants	
8	Actual planting date	
9	Row feet planted	
10	Harvest date	
11	Harvest amount	
12	Notes	

PLANTING DATES WORKSHEET

Ideal earliest and latest planting dates can vary widely depending on your latitude, local microclimate, and days to maturity of the variety you're growing. These dates assume you're not using season-extension infrastructure (if you're using a high tunnel or row covers, earlier first plantings and later last plantings are possible). The dates here are general guidelines; check with other growers in your area and don't be afraid to experiment to fine-tune your planting dates.

Perennial herbs can be planted in spring and early summer in colder climates, and in fall in areas with mild winters (USDA Zones 7 or 8 or higher). In general, perennial garden fruits can be planted in spring and

ANNUAL VEGETABLES AND HERBS

CROP	RECOMMENDED FIRST OUTDOOR TRANSPLANTING	MY DATE IS	RECOMMENDED FIRST OUTDOOR DIRECT SEEDING	MY DATE IS	RECOMMENDED LAST PLANTING DATE
Arugula	Not recommended	—	4 weeks before last frost		3–4 weeks before first frost (direct seed)
Basil	3 weeks after last frost		3 weeks after last frost		10–12 weeks before first frost (transplant)
Beans, edible soy (edamame)	2–8 weeks after last frost		On last frost date		12–16 weeks before first frost (direct seed)
Beans, fava (broad)	2–8 weeks after last frost		10 weeks before last frost		Varies with climate
Beans, lima	2–8 weeks after last frost		When soil temperature reaches 75°F		10 weeks before first frost (direct seed)
Beans, shell (fresh or dried)	2–8 weeks after last frost		On last frost date		10 weeks before first frost (direct seed)
Beans, snap (bush or pole)	2–8 weeks after last frost		On last frost date		10 weeks before first frost (direct seed)
Beets	Not recommended	—	On last frost date		12 weeks before first frost (direct seed)
Broccoli	2 weeks before last frost		2 weeks after last frost		10–12 weeks before first frost (transplant)
Broccoli raab	2 weeks before last frost		2 weeks before last frost		4–6 weeks before first frost (transplant)
Brussels sprouts	On last frost date		2 weeks after last frost		12 weeks before first frost (transplant)

early summer in colder climates, and in fall in areas with mild winters (Zones 7 or higher). The exception is strawberries, which can be planted in spring or fall in a variety of climates. Fruit and nut trees can be planted in spring and early summer in colder climates, and in fall in areas with mild winters (Zones 7 or higher). And generally, perennial vines can be planted in spring and early summer in colder climates, and in fall in areas with mild winters (Zones 7 or higher).

MY LAST FROST DATE IS: _____ MY FIRST FROST DATE IS: _____

MY DATE IS	SOIL TEMP. RANGE FOR GERMINATION (°F)	OPTIMUM GERMINATION TEMP. (°F)	APPROXIMATE GROW TIME FROM NURSERY SEEDING TO TRANSPLANTING	NOTES
	45–82	67	N/A	Easy and efficient to direct seed
	70–90	80	6–8 weeks	Transplanting preferred in most climates
	50–95	85	2–4 weeks	Direct seeding preferred
	40–75	68	2–3 weeks	Direct seeding preferred
	75–95	85	2–3 weeks	Direct seeding preferred
	60–95	80	2–3 weeks	Direct seeding preferred
	60–95	80	2–3 weeks	Direct seeding preferred
	50–80	77	4–6 weeks	Transplanting adversely affects growth
	45–90	85	4 weeks	Transplanting preferred in most climates
	45–85	77	3–4 weeks	Transplant or direct seed
	50–90	86	4 weeks	Transplanting preferred in most climates

CROP	RECOMMENDED FIRST OUTDOOR TRANSPLANTING	MY DATE IS	RECOMMENDED FIRST OUTDOOR DIRECT SEEDING	MY DATE IS	RECOMMENDED LAST PLANTING DATE
Cabbage	2 weeks before last frost		2 weeks before last frost		12 weeks before first frost (transplant)
Cabbage, Chinese	On last frost date		2 weeks before last frost		6–8 weeks before first frost (transplant)
Carrots	Not recommended	—	4–6 weeks before last frost		12–18 weeks before first frost (direct seed)
Cauliflower	2 weeks before last frost		2 weeks before last frost		12 weeks before first frost (transplant)
Celeriac	2 weeks after last frost		2 weeks after last frost		16 weeks before first frost (transplant)
Celery	2 weeks after last frost		2 weeks after last frost		16 weeks before first frost (transplant)
Chard, Swiss	2 weeks before last frost		On last frost date		8 weeks before first frost (transplant)
Cilantro	Not recommended	—	On last frost date		4–6 weeks before first frost (direct seed)
Collards	2–4 weeks before last frost		4 weeks before last frost		8 weeks before first frost (transplant)
Corn, sweet	2 weeks after last frost		When soil temperature is above 65°F		Varies with climate; generally, early summer
Cucumbers	4 weeks after last frost		3–4 weeks after last frost		12 weeks before first frost (transplant)
Dill	On last frost date		On last frost date		4–6 weeks before first frost for leaf harvest (direct seed)
Eggplant	4 weeks after last frost		4 weeks after last frost		16 weeks before first frost (transplant)
Endive	2–4 weeks before last frost		On last frost date		4–6 weeks before first frost (transplant)
Fennel, bulbing	2 weeks before last frost		2–4 weeks after last frost		12 weeks before first frost (transplant)

MY DATE IS	SOIL TEMP. RANGE FOR GERMINATION (°F)	OPTIMUM GERMINATION TEMP. (°F)	APPROXIMATE GROW TIME FROM NURSERY SEEDING TO TRANSPLANTING	NOTES
	50–95	88	4 weeks	Transplanting preferred in most climates
	52–95	87	4 weeks	Transplanting preferred in most climates
	41–98	77	N/A	Direct seeding preferred; transplanting generally produces poor yields
	45–90	85	4–6 weeks	Transplanting preferred in most climates
	41–100	75	10–12 weeks	Transplanting preferred in most climates
	41–100	75	10–12 weeks	Transplanting preferred in most climates
	41–95	86	5–6 weeks	Transplanting preferred in most climates
	50–80	70	N/A	Direct seeding preferred; transplanted cilantro is prone to bolting
	55–95	88	4–6 weeks	Transplanting preferred in most climates
	60–100	86	2–4 weeks	Commonly direct seeded, but transplanting is very effective for small plots
	60–95	86	3–4 weeks	Transplanting best for early crops; direct seeding effective when soil has warmed in early summer
	50–80	70	4–6 weeks	Most efficient to direct seed
	65–100	90	8–10 weeks	Transplanting preferred in most climates
	41–85	68	3–4 weeks	Transplant or direct seed
	50–100	77	4–6 weeks	Transplanting preferred in most climates

CROP	RECOMMENDED FIRST OUTDOOR TRANSPLANTING	MY DATE IS	RECOMMENDED FIRST OUTDOOR DIRECT SEEDING	MY DATE IS	RECOMMENDED LAST PLANTING DATE
Garlic	N/A	—	N/A	—	One planting
Kale	2–4 weeks before last frost		2 weeks before last frost		8 weeks before first frost (transplant)
Kohlrabi	2 weeks before last frost		2 weeks before last frost		8 weeks before first frost (transplant or direct seed)
Leeks	On last frost date		2 weeks before last frost		12–16 weeks before first frost (transplant)
Lettuce, baby mix	N/A	—	2–4 weeks before last frost		4 weeks before first frost (direct seed)
Lettuce, head	2–4 weeks before last frost		2–4 weeks before last frost		4 weeks before first frost (transplant)
Mâche	As early as soil can be worked		As early as soil can be worked		2–4 weeks before first frost (direct seed or transplant)
Melon, cantaloupe, honeydew	4 weeks after last frost		When soil temperature reaches 70°F		16 weeks before first frost (transplant)
Mustard greens	4 weeks before last frost		4 weeks before last frost		2–4 weeks before first frost (direct seed or transplant)
Okra	4–6 weeks after last frost		When soil temperature reaches 65°F		12–16 weeks before first frost (transplant)
Onions, bulb	2 weeks before last frost		6 weeks before last frost		Variable depending on type
Pak choi	On last frost date		On last frost date		4–8 weeks before first frost (direct seed or transplant)
Parsley	2 weeks before last frost		2 weeks before last frost		8–12 weeks before first frost (transplant or direct seed)
Parsnips	Not recommended	—	As early as soil can be worked		20–24 weeks before first frost (direct seed)

MY DATE IS	SOIL TEMP. RANGE FOR GERMINATION (°F)	OPTIMUM GERMINATION TEMP. (°F)	APPROXIMATE GROW TIME FROM NURSERY SEEDING TO TRANSPLANTING	NOTES
	N/A	N/A	N/A	Plant cloves in late fall (Oct.–Nov. in most of North America) for harvest the following summer
	52–100	90	4–6 weeks	Transplanting preferred in most climates
	48–86	77	3–4 weeks	Transplant or direct seed
	42–94	77	6–8 weeks	Transplanting preferred in most climates
	41–85	68	N/A	Direct seed
	41–85	68	3–4 weeks	Transplanting preferred in most climates
	41–68	63	4–6 weeks	Transplant or direct seed
	68–100	90	3–4 weeks	Transplanting best for early crops; direct seeding effective when soil has warmed in early summer
	45–86	77	3–4 weeks	Most efficient to direct seed
	60–95	85	4–5 weeks	Transplanting preferred in most climates; is a heat lover and grows best in warm climates
	45–95	77–86	8–10 weeks	Transplanting preferred in most climates
	45–85	75	3–4 weeks	Transplant or direct seed
	50–85	75	6–8 weeks	Transplants easier to manage due to slow germination
	40–80	65	N/A	Direct seed

CROP	RECOMMENDED FIRST OUTDOOR TRANSPLANTING	MY DATE IS	RECOMMENDED FIRST OUTDOOR DIRECT SEEDING	MY DATE IS	RECOMMENDED LAST PLANTING DATE
Peanut	When soil temperature reaches 65°F		When soil temperature reaches 65+°F		16–20 weeks before first frost
Peas, shelling	4 weeks before last frost		As early as soil can be worked		Generally, plant in early–mid spring; fall crops are possible; direct seed about 8 weeks before first frost
Peas, snap	4 weeks before last frost		As early as soil can be worked		Generally, plant in early–mid spring; fall crops are possible; direct seed about 8 weeks before first frost
Peppers, hot	2 weeks after last frost		Not recommended	—	16 weeks before first frost (transplant)
Peppers, sweet	2 weeks after last frost		Not recommended	—	16 weeks before first frost (transplant)
Potatoes	N/A	—	4 weeks before last frost		N/A; one planting in spring
Radicchio	2 weeks before last frost		2 weeks before last frost		Variable; generally, plant in early spring and late summer for fall production
Radishes	Not recommended	—	4 weeks before last frost		4 weeks before first frost (direct seed)
Rutabagas	Not recommended	—	12–14 weeks before first frost		12–14 weeks before first frost (direct seed)
Scallions	6 weeks before last frost		6 weeks before last frost		8 weeks before first frost (transplant or direct seed)
Spinach	4 weeks before last frost		As early as soil can be worked		2–4 weeks before first frost (direct seed or transplant)
Squash, gourds	2 weeks after last frost		When soil temperature has reached 70°F		Variable depending on type
Squash, pumpkins	2 weeks after last frost		When soil temperature has reached 70°F		Variable depending on type

MY DATE IS	SOIL TEMP. RANGE FOR GERMINATION (°F)	OPTIMUM GERMINATION TEMP. (°F)	APPROXIMATE GROW TIME FROM NURSERY SEEDING TO TRANSPLANTING	NOTES
	65–100	86–91	4–6 weeks	Direct seed
	40–75	75	2–3 weeks	Transplant or direct seed
	40–75	75	2–3 weeks	Transplant or direct seed
	68–95	85	6–8 weeks	Transplanting preferred in most climates
	68–95	85	6–8 weeks	Transplanting preferred in most climates
	N/A	N/A	N/A	Plant tubers in early spring
	40–85	75	3–4 weeks	Transplant or direct seed
	45–90	85	N/A	Direct seed
	65–85	77	N/A	Direct seed
	45–95	75	6–8 weeks	Transplant or direct seed
	45–75	70	3–4 weeks	Transplant or direct seed
	60–95	86	3–4 weeks	Transplanting best for early crops; direct seeding effective when soil has warmed in early summer
	60–95	86	3–4 weeks	Transplanting best for early crops; direct seeding effective when soil has warmed in early summer

CROP	RECOMMENDED FIRST OUTDOOR TRANSPLANTING	MY DATE IS	RECOMMENDED FIRST OUTDOOR DIRECT SEEDING	MY DATE IS	RECOMMENDED LAST PLANTING DATE
Squash, summer	2 weeks after last frost		When soil temperature has reached 70°F		12 weeks before first frost (transplant)
Squash, winter	2 weeks after last frost		When soil temperature has reached 70°F		Variable depending on type
Sweet potatoes	After last frost, when soil temperature is above 60°F		N/A	—	N/A
Tomatillos	2 weeks after last frost		Not recommended	—	Variable; generally, 12–16 weeks before first frost
Tomatoes	2 weeks after last frost		Not recommended	—	Variable depending on type
Turnips	Not recommended	—	2 weeks before last frost		6–8 weeks before first frost (direct seed)
Watermelon	4 weeks after last frost		When soil temperature has reached 70°F		Variable; generally, 16 weeks before first frost

PERENNIAL VEGETABLES

CROP	RECOMMENDED FIRST OUTDOOR TRANSPLANTING	MY DATE IS	NOTES
Artichoke	2 weeks after last frost		Use transplants
Asparagus	Spring		Use transplants or crowns
Cardoon	Spring		Use transplants or root pieces
Jerusalem artichoke, sunchoke	Fall		Grow from tubers
Rhubarb	Spring or fall		Use transplants or root pieces

MY DATE IS	SOIL TEMP. RANGE FOR GERMINATION (°F)	OPTIMUM GERMINATION TEMP. (°F)	APPROXIMATE GROW TIME FROM NURSERY SEEDING TO TRANSPLANTING	NOTES
	60–95	86	3–4 weeks	Transplanting best for early crops; direct seeding effective when soil has warmed in early summer
	60–95	86	3–4 weeks	Transplanting best for early crops; direct seeding effective when soil has warmed in early summer
	N/A	N/A	N/A	Sweet potatoes usually grown from slips, or bare-root transplants propagated from the tubers of the plant
	60–90	85	4–5 weeks	Transplanting preferred in most climates
	60–90	85	6–8 weeks	Transplanting preferred in most climates
	50–86	77	N/A	Direct seed
	68–100	95	4 weeks	Transplanting best for early crops; direct seeding effective when soil has warmed in early summer

GARDEN PLANNING CHART

ANNUAL VEGETABLES & HERBS	CROP FAMILY	A.K.A.	SCIENTIFIC NAME	CROP LIFE SPAN	DAYS TO MATURITY	SPACING	DIRECT SEED (DS) OR TRANSPLANT (TP)
Arugula	Brassicaceae	Brassicas	*Eruca vesicaria*	Short season	30–60 from DS	Thickly seeded in rows, approx. 30 plants per row foot	DS
Basil	Lamiaceae	Mints	*Ocimum basilicum*	Half season	30–60 from TP	6"–12"	TP
Beans, edible soy (edamame)	Fabaceae	Legumes	*Glycine max*	Long season	75–90 from DS	1.5"–2"	DS, TP
Beans, fava (broad)	Fabaceae	Legumes	*Vicia faba*	Long season	75–120 from DS	4"–6"	DS, TP
Beans, lima	Fabaceae	Legumes	*Phaseolus lunatus*	Half season	60–90 from DS	3"	DS, TP
Beans, shell (fresh or dried)	Fabaceae	Legumes	*Phaseolus vulgaris*	Long season	60–70 fresh, 80–110 dried, from DS	3"–4"	DS, TP
Beans, snap (bush types)	Fabaceae	Legumes	*Phaseolus vulgaris*	Half season	50–60 from DS	2"–3"	DS, TP
Beans, snap (pole types)	Fabaceae	Legumes	*Phaseolus vulgaris*	Half season	50–60 from DS	2"–3"	DS, TP
Beets	Amaranthaceae	Amaranths, chenopods	*Beta vulgaris*	Half season	50–60 from DS	4"	DS, TP
Broccoli	Brassicaceae	Brassicas	*Brassica oleracea*	Half season	50–60 from TP	12"–18"	TP
Broccoli raab	Brassicaceae	Brassicas	*Brassica rapa*	Short season	35–45 from TP	1"–2"	DS, TP
Brussels sprouts	Brassicaceae	Brassicas	*Brassica oleracea*	Long season	90–120 from TP	18"	TP
Cabbage	Brassicaceae	Brassicas	*Brassica oleracea*	Half season	60–90 from TP	12"–18"	TP
Cabbage, Chinese	Brassicaceae	Brassicas	*Brassica rapa*	Half season	50–60 from TP	12"–18"	TP

TOLERATE ROOT DISTURBANCE AS TRANSPLANTS	SUCCESSIONS	CROP HEIGHT	TRELLISING	FERTILITY NEEDS	NEED SUPPLEMENTAL FERTILIZING	FLOWER TYPES	SELF-POLLINATING
N/A	Weekly	6"–8"		Low	No	Hermaphroditic	Yes
Yes	1–3 plantings	24"	Hilling, single stake	Low	No	Hermaphroditic	Yes
No	1–3 plantings	24"	Hilling, single stake	Low	No	Hermaphroditic	Yes
No	1–3 plantings in spring, fall planting in areas with winter lows above 10°F	36"	Hilling, single stake	Low	No	Hermaphroditic	Yes
No	1–3 plantings	24"		Low	No	Hermaphroditic	Yes
No	1–3 plantings	24" bush / 6' pole	Hilling, single stake for bush types; cage, fence, ladder, net, string, tripod for pole types	Low	No	Hermaphroditic	Yes
No	Every 2–3 weeks	24"	Hilling, single stake	Low	No	Hermaphroditic	Yes
No	1–2 plantings	6'	Cage, fence, ladder, net, string, tripod	Low	No	Hermaphroditic	Yes
No	Every 2–3 weeks	6"–8"		High	Yes	Hermaphroditic	Yes
Yes	Every 2–3 weeks	24"	Hilling, single stake	High	Yes	Hermaphroditic	No
Yes	Every 1–2 weeks	24"–36"	Hilling, single stake	High	Yes	Hermaphroditic	Yes
Yes	1–2 plantings in spring	36"–42"	Hilling, single stake	High	Yes	Hermaphroditic	No
Yes	Every 2–3 weeks	12"		High	Yes	Hermaphroditic	No
Yes	Every 2–3 weeks	12"		High	Yes	Hermaphroditic	No

ANNUAL VEGETABLES & HERBS	CROP FAMILY	A.K.A.	SCIENTIFIC NAME	CROP LIFE SPAN	DAYS TO MATURITY	SPACING	DIRECT SEED (DS) OR TRANSPLANT (TP)
Carrots	Apiaceae	Umbels	*Daucus carota*	Half season	50–80 from DS	2"	DS
Cauliflower	Brassicaceae	Brassicas	*Brassica oleracea*	Half season	50–80 from TP	12"–18"	TP
Celeriac	Apiaceae	Umbels	*Apium graveolens*	Long season	85–110 from TP	8"	TP
Celery	Apiaceae	Umbels	*Apium graveolens*	Long season	85–95 from TP	8"–12"	TP
Chard, Swiss	Amaranthaceae	Amaranths, chenopods	*Beta vulgaris*	Half season	50–60 from DS	8"–12"	TP
Cilantro	Apiaceae	Umbels	*Coriandrum sativum*	Half season	50–60 from DS	Sow thickly in rows, 12–24 plants per row foot	DS
Collards	Brassicaceae	Brassicas	*Brassica oleracea*	Half season	50–60 from TP	8"–12"	TP
Corn, sweet	Poaceae	Grasses	*Zea mays*	Long season	65–85 from DS	8"–12"	DS, TP
Cucumbers	Cucurbitaceae	Cucurbits	*Cucumis sativus*	Half season	50–60 from TP	12" if trellised, 5 sq. ft. per plant if sprawling	DS, TP
Dill	Apiaceae	Umbels	*Anethum graveolens*	Short season	40–60 from DS	9"	DS
Eggplant	Solanaceae	Nightshades, solanums	*Solanum melongena*	Half season	55–75 from TP	12"	TP
Endive	Asteraceae	Daisies	*Cichorium endivia*	Short season	45–60 from TP	8"–12"	TP
Fennel, bulbing	Apiaceae	Umbels	*Foeniculum vulgare*	Long season	75–85 from TP	6"–12"	TP
Garlic	Amaryllidaceae	Alliums	*Allium sativum*	Long season	210+ (fall planted from cloves)	6"–8"	Planted from cloves
Kale	Brassicaceae	Brassicas	*Brassica oleracea*	Half season	50–60 from TP	8"–12"	TP
Kohlrabi	Brassicaceae	Brassicas	*Brassica oleracea*	Half season	40–80 from TP	4"–6"	TP
Leeks	Amaryllidaceae	Alliums	*Allium porrum*	Long season	70–120 from TP	6"	TP

TOLERATE ROOT DISTURBANCE AS TRANSPLANTS	SUCCESSIONS	CROP HEIGHT	TRELLISING	FERTILITY NEEDS	NEED SUPPLEMENTAL FERTILIZING	FLOWER TYPES	SELF-POLLINATING
N/A	Every 2–3 weeks	6"–8"		High	Yes	Hermaphroditic	Yes
Yes	Every 2–3 weeks	12"–16"		High	Yes	Hermaphroditic	No
Yes	1–2 plantings in spring	6"–8"		Medium	No	Hermaphroditic	Yes
Yes	1–2 plantings in spring	12"–16"		Medium	No	Hermaphroditic	Yes
Yes	Every 4 weeks	24"–36"		Medium	No	Hermaphroditic	Yes
N/A	Weekly	6"–8"		Low	No	Hermaphroditic	Yes
Yes	Every 4 weeks	36"–42"	Hilling, single stake	High	Yes	Hermaphroditic	Yes
No	Weekly	5'–8'		High	Yes	Monoecious	Yes
No	Every 4 weeks	Sprawling or 5'–8' on trellis	Cage, fence, ladder, net, string, tripod	High	Yes	Monoecious	Variety dependent
No	Weekly	4'–5'		Low	No	Hermaphroditic	Yes
Yes	1–2 plantings in spring	36"	Hilling, single stake	High	Yes	Hermaphroditic	Yes
Yes	Weekly in spring and fall	12"		Low	No	Hermaphroditic	Yes
Yes	Weekly	12"		Low	No	Hermaphroditic	Yes
N/A	One planting in fall	12"–16"		Medium	Yes	Hermaphroditic	No
Yes	Every 4 weeks	36"–42"	Hilling, single stake	High	Yes	Hermaphroditic	Yes
Yes	Weekly	12"		Low	No	Hermaphroditic	Yes
Yes	1–2 plantings in spring	12"–24"		Medium	Yes	Hermaphroditic	Yes

ANNUAL VEGETABLES & HERBS	CROP FAMILY	A.K.A.	SCIENTIFIC NAME	CROP LIFE SPAN	DAYS TO MATURITY	SPACING	DIRECT SEED (DS) OR TRANSPLANT (TP)
Lettuce, baby mix	Asteraceae	Daisies	*Lactuca sativa*	Short season	25–40 from DS	6"–12"	DS
Lettuce, head	Asteraceae	Daisies	*Lactuca sativa*	Short season	30–45 from TP	Thickly seeded in rows, approx. 30 plants per row foot	TP
Mâche	Caprifoliaceae	Honeysuckles	*Valerianella locusta*	Half season	50–60 from DS	1"	DS
Melon, cantaloupe, honeydew	Cucurbitaceae	Cucurbits	*Cucumis melo*	Long season	70–90 from TP	2'–3'	DS, TP
Mustard greens	Brassicaceae	Brassicas	*Brassica* spp.	Short season	30–40 baby, 40–50 full size, from DS	For baby greens, sow thickly in rows, approx. 30 plants per row foot; for full-size greens, space at approx. 1"	DS
Okra	Malvaceae	Mallows	*Abelmoschus esculentus*	Long season	50–60 from TP	12"–18"	TP
Onions, bulb	Amaryllidaceae	Alliums	*Allium cepa*	Long season	90–110 from TP	4"–6"	TP
Pak choi	Brassicaceae	Brassicas	*Brassica rapa*	Short season	45–60 from TP	For baby size, 2"–4", for full size, 6"–12"	TP
Parsley	Apiaceae	Umbels	*Petroselinum crispum*	Long season	30–60 from TP	8"–12"	TP
Parsnips	Apiaceae	Umbels	*Pastinaca sativa*	Long season	100–120 from DS	2"	DS
Peanut	Fabaceae	Legumes	*Arachis hypogaea*	Long season	90–150 from DS	10"	DS
Peas, shelling	Fabaceae	Legumes	*Pisum sativum*	Half season	50–65 from DS	1"–2"	DS, TP
Peas, snap	Fabaceae	Legumes	*Pisum sativum*	Half season	50–65 from DS	1"–2"	DS, TP

TOLERATE ROOT DISTURBANCE AS TRANSPLANTS	SUCCESSIONS	CROP HEIGHT	TRELLISING	FERTILITY NEEDS	NEED SUPPLEMENTAL FERTILIZING	FLOWER TYPES	SELF-POLLINATING
N/A	Weekly	8"–12"		Low	No	Hermaphroditic	Yes
Yes	Weekly	6"–8"		Low	No	Hermaphroditic	Yes
Yes	Every 1–2 weeks in early spring and fall	6"		Low	No	Hermaphroditic	Yes
No	1–3 plantings	Sprawling or 6'–10' on trellis	Cage, fence, ladder, net, string, tripod	High	Yes	Monoecious	Yes
Yes	Every 1–2 weeks	6"–18"		Low	No	Hermaphroditic	Yes
Yes	1–3 plantings	36"	Hilling, single stake	High	Yes	Hermaphroditic	Yes
Yes	1 planting	12"–18"		Medium	Yes	Hermaphroditic	Yes
Yes	Every 1–2 weeks	12"–18"		Low	No	Hermaphroditic	Yes
Yes	1–3 plantings	12"		Low	No	Hermaphroditic	Yes
N/A	1 planting	6"–8"		Medium	Yes	Hermaphroditic	Yes
No	1 planting	12"–18"		High	Yes	Hermaphroditic	Yes
No	1–3 plantings in spring, 1–2 in fall	24" bush/ 6' pole	Cage, fence, ladder, net, string, tripod	Low	No	Hermaphroditic	Yes
No	1–3 plantings in spring, 1–2 in fall	24" bush/ 6' pole	Cage, fence, ladder, net, string, tripod	Low	No	Hermaphroditic	Yes

ANNUAL VEGETABLES & HERBS	CROP FAMILY	A.K.A.	SCIENTIFIC NAME	CROP LIFE SPAN	DAYS TO MATURITY	SPACING	DIRECT SEED (DS) OR TRANSPLANT (TP)
Peppers, hot	Solanaceae	Nightshades, Solanums	*Capsicum annuum*	Long season	70–120 from TP	12"	TP
Peppers, sweet	Solanaceae	Nightshades, Solanums	*Capsicum annuum*	Long season	70–120 from TP	12"	TP
Potatoes	Solanaceae	Nightshades, Solanums	*Solanum tuberosum*	Long season	70–120 (from tubers)	12"	From tubers
Radicchio	Asteraceae	Daisies	*Cichorium intybus*	Half season	55–70 from TP	10"–12"	TP
Radishes	Brassicaceae	Brassicas	*Rhaphanus sativus*	Short season	30–60 from DS	1"–2"	DS
Rutabagas	Brassicaceae	Brassicas	*Brassica napobrassica*	Long season	90–100 from DS	6"–8"	DS
Scallions	Amaryllidaceae	Alliums	*Allium wakegi*	Half season	60–70 from DS	1"	DS, TP
Spinach	Amaranthaceae	Alliums	*Spinacia oleracea*	Short season	30–40 from DS	1"–2"	DS, TP
Squash, gourds	Cucurbitaceae	Cucurbits	*Cucurbita* spp.	Long season	90–130 from TP	2'–3'	DS, TP
Squash, pumpkins	Cucurbitaceae	Cucurbits	*Cucurbita pepo*	Long season	80–120 from TP	2'–3'	DS, TP
Squash, summer	Cucurbitaceae	Cucurbits	*Cucurbita pepo, Cucurbita maxima*	Half season	45–60 from TP	2'–3"	DS, TP
Squash, winter	Cucurbitaceae	Cucurbits	*Cucurbita* spp.	Long season	80–120 from TP	2'–3'	DS, TP
Sweet potatoes	Convolvulaceae	Morning Glories	*Ipomoea batatas*	Long season	90–100 (from slips)	12"–18"	From slips
Tomatillos	Solanaceae	Nightshades, solanums	*Physalis ixocarpa*	Long season	55–75 from TP	18"–24"	TP
Tomatoes	Solanaceae	Nightshades, solanums	*Lycopersicon esculentum*	Long season	55–85 from TP	18"–24"	TP
Turnips	Brassicaceae	Brassicas	*Brassica rapa*	Short season	30–60 from DS	4"–6"	DS
Watermelon	Cucurbitaceae	Cucurbits	*Citrullus lanatus*	Long season	70–90 from TP	2'–3'	DS, TP

TOLERATE ROOT DISTURBANCE AS TRANSPLANTS	SUCCESSIONS	CROP HEIGHT	TRELLISING	FERTILITY NEEDS	NEED SUPPLEMENTAL FERTILIZING	FLOWER TYPES	SELF-POLLINATING
Yes	1–3 plantings	24"–36"	Hilling, single stake	High	Yes	Hermaphroditic	Yes
Yes	1–3 plantings	24"–36"	Hilling, single stake	High	Yes	Hermaphroditic	Yes
N/A	1 planting	12"–24"	Hilling, single stake	High	Yes	Hermaphroditic	Yes
Yes	Every 1–2 weeks in spring and fall	12"		Medium	No	Hermaphroditic	Yes
N/A	Weekly	6"		Low	No	Hermaphroditic	Yes
Yes	1–3 plantings	12"–18"		Medium	Yes	Hermaphroditic	Yes
Yes	Every 1–2 weeks	12"		Low	No	Hermaphroditic	Yes
Yes	Every 1–2 weeks in spring and fall	6"–12"		Low	No	Dioecious	Yes
No	1–3 plantings	Sprawling or 6'–10' on trellis	Cage, fence, ladder, net, string, tripod	High	Yes	Monoecious	Yes
No	1–3 plantings	Sprawling or 6'–10' on trellis	Cage, fence, ladder, net, string, tripod	High	Yes	Monoecious	Yes
No	1–3 plantings	36"		High	Yes	Monoecious	Yes
No	1–3 plantings	Sprawling or 6'–10' on trellis	Cage, fence, ladder, net, string, tripod	High	Yes	Monoecious	Yes
Yes	1 planting	12"		High	Yes	Hermaphroditic	Yes
Yes	1–3 plantings	36"	Cage, single stake	High	Yes	Hermaphroditic	No
Yes	1–3 plantings	36" determinate / 6'–10' indeterminate	Cage, fence, ladder, net, string, tripod	High	Yes	Hermaphroditic	Yes
N/A	Every 1–2 weeks	12"		Medium	Yes	Hermaphroditic	Yes
No	1–3 plantings	Sprawling or 6'–10' on trellis	Cage, fence, ladder, net, string, tripod	High	Yes	Monoecious	Yes

RESOURCES

suggested reading

BOOKS

American Horticultural Society Pests and Diseases: The Complete Guide to Preventing, Identifying, and Treating Plant Problems. Pippa Greenwood, Andrew Halstead, A.R. Chase, and Daniel Gilrein. New York: DK, 2000.

The Ball Complete Book of Home Preserving. Judi Kingry, Lauren Devine, and Sarah Page. Toronto: Robert Rose, 2020.

Ball Perennial Manual: Propagation and Production. Jim Nau. Greeneville, OH: Ball, 1996.

Bugs, Slugs, & Other Thugs: Controlling Garden Pests Organically. Rhonda Massingham Hart. North Adams, MA: Storey, 1991.

Carrots Love Tomatoes: Secrets of Companion Planting for Successful Gardening. Louise Riotte. North Adams, MA: Storey, 1998.

Cover Crop Gardening: Soil Enrichment with Green Manures. Editors of Storey Publishing. Storey Country Wisdom Bulletin A-05. North Adams, MA: Storey, 1983.

Designing and Maintaining Your Edible Landscape Naturally. Robert Kourik. White River Junction, VT: Chelsea Green, 2005.

Drip Irrigation for Every Landscape and All Climates, 2nd ed. Robert Kourik. Occidental, CA: Metamorphic, 2009.

Easy Composters You Can Build. Nick Noyes. Storey Country Wisdom Bulletin A-139. North Adams, MA: Storey, 1995.

Edible Wild Plants: A North American Field Guide to Over 200 Natural Foods. Thomas Elias and Peter Dykeman. New York: Sterling, 2009.

Food Grown Right, In Your Backyard: A Beginner's Guide to Growing Crops at Home. Colin McCrate and Brad Halm. Seattle, WA: Skipstone, 2012.

Four-Season Harvest: Organic Vegetables from Your Home Garden All Year Long. Eliot Coleman. White River Junction, VT: Chelsea Green, 1999.

Fresh & Fermented: 85 Delicious Ways to Make Fermented Carrots, Kraut, and Kimchi Part of Every Meal. Julie O'Brien and Richard J. Climenhage. Seattle: Sasquatch, 2014.

Give Peas a Chance! Organic Gardening Cartoon-Science. Peter Barbarow. Happy Camp, CA: Naturegraph, 1990.

Growing Great Garlic: The Definitive Guide for Organic Gardeners and Small Farmers. Ron L. Engeland. Okanogan, WA: Filaree, 1991.

Lasagna Gardening: A New Layering System For Bountiful Gardens: No Digging, No Tilling, No Weeding, No Kidding! Patricia Lanza. Emmaus, PA: Rodale, 1998.

Managing Cover Crops Profitably. SARE Outreach. Andy Clark, ed. College Park, MD: SARE Outreach, 2007.

On-Farm Composting Handbook. Robert Rynk, ed. Ithaca, NY: Plant and Life Sciences, 1992.

Putting Food By, 5th ed. Ruth Hertzberg, Janet Greene, and Beatrice Vaughan. New York: Plume, 2010.

Seed to Seed: Seed Saving and Growing Techniques for Vegetable Gardeners, 2nd ed. Suzanne Ashworth. Decorah, IA: Seed Saver's Exchange, 2002.

Sharing the Harvest: A Citizen's Guide to Community Supported Agriculture, 2nd ed. Elizabeth Henderson and Robyn Van En. White River Junction, VT: Chelsea Green, 2007.

Small-Scale Grain Raising: An Organic Guide to Growing, Processing, and Using Nutritious Whole Grains for Home Gardeners and Local Farmers. Gene Logsdon. White River Junction, VT: Chelsea Green, 1977.

Storey's Guide to Raising Chickens, 3rd ed. Gail Damerow. North Adams, MA: Storey, 2010.

The Complete Book of Edible Landscaping. Rosalind Creasy. San Francisco: Sierra Club Books, 1982.

The Intelligent Gardener: Growing Nutrient-Dense Food. Steve Soloman. Gabriola Island, BC: New Society, 2012.

The Market Gardener: A Successful Grower's Handbook for Small-Scale Organic Farming. Jean-Martin Fortier. Gabriola Island, BC: New Society, 2014.

The New Organic Grower: A Master's Manual of Tools and Techniques for the Home and Market Gardener, 2nd ed. Eliot Coleman. White River Junction, VT: Chelsea Green, 1995.

The Organic Farmer's Business Handbook: A Complete Guide to Managing Finances, Crops, and Staff—and Making a Profit. Richard Wiswall. White River Junction, VT: Chelsea Green, 2009.

The Rodale Guide to Composting: Easy Methods for Every Gardener. Jerry Minnich and Marjorie Hunt. Emmaus, PA: Rodale, 1979.

The Self-Sufficient Suburban Gardener. Jeff Ball. Emmaus, PA: Rodale, 1983.

The Soul of Soil: A Soil-Building Guide for Master Gardeners and Farmers, 4th ed. Joe Smillie and Grace Gershuny. White River Junction, VT: Chelsea Green, 1999.

The Sprouting Book: How to Grow and Use Sprouts to Maximize Your Health and Vitality. Ann Wigmore. Garden City, NY: Avery, 1986.

The New Vegetable Growers Handbook: A Users Manual for the Organic Vegetable Garden. Frank Tozer. Santa Cruz, CA: Green Man, 2013.

The Winter Harvest Handbook: Year Round Vegetable Production Using Deep-Organic Techniques and Unheated Greenhouses. Eliot Coleman. White River Junction, VT: Chelsea Green, 2009.

Wholesale Success: A Farmer's Guide to Food Safety, Selling, Postharvest Handling, and Packing Produce, 3rd ed. Jim Slama and Atina Diffley, eds. Oak Park, IL: FamilyFarmed.org, 2012.

Wild Fermentation: The Flavor, Nutrition, and Craft of Live-Culture Foods. Sandor Ellix Katz. White River Junction, VT: Chelsea Green, 2003.

Worms Eat My Garbage, 35th Anniversary Edition: How to Set Up and Maintain a Worm Composting System. Mary Appelhof and Joanne Olszewski. North Adams, MA: Storey, 2019.

Your Backyard Herb Garden: A Gardener's Guide to Growing Over 50 Herbs Plus How to Use Them In Cooking, Crafts, Companion Planting and More. Miranda Smith. Emmaus, PA: Rodale, 1999.

MAGAZINES

Growing for Market
www.growingformarket.com

Mother Earth News
www.motherearthnews.com

online resources & supplies

SITE MAPPING

Google Earth
www.google.com/earth

Google Maps
www.google.com/maps

Inkscape
www.inkscape.org

SketchUp
www.sketchup.com

SOIL TESTING LABS

A & L Western Laboratories
www.al-labs-west.com

University of Massachusetts Amherst Soil and Plant Nutrition Testing Laboratory
https://soiltest.umass.edu

SEED SUPPLIERS

We've found the following seed companies carry quality seed in a range of quantities that are useful for serious home gardeners and small-scale farmers. They also provide very helpful resources for growers on their website and useful information on their seed packets.

Fedco Seeds
www.fedcoseeds.com

High Mowing Organic Seeds
www.highmowingseeds.com

Johnny's Selected Seeds
www.johnnysseeds.com

TOOL SUPPLY & DIY INFORMATION

CoolBot
www.storeitcold.com
DIY walk-in cooler

FarmHack
https://farmhack.org
Online resource for DIY farm tool builders

Farmtek
www.farmtek.com
Season-extension equipment

Flame Engineering
www.flameengineering.com
Flame weeding supplies

Johnny's Selected Seeds
www.johnnysseeds.com
Season-extension equipment, tools, occultation tarps

Neversink Tools
www.neversinktools.com
Tools, occultation tarps, flame weeders

ORGANIC PEST & DISEASE CONTROL & FERTILIZERS

ARBICO Organics
www.arbico-organics.com

Dr. Earth
https://drearth.com

Planet Natural
www.planetnatural.com

IRRIGATION SUPPLIES

The Drip Store
www.dripirrigation.com

DripWorks
www.dripworksusa.com

Irrigation King
www.irrigationking.com

Senninger Irrigation
www.senninger.com

NONTOXIC WOOD PRESERVATIVES FOR RAISED BEDS

Internal Wood Stabilizer from TimberPro Coatings
https://timberprocoatingsusa.com

LifeTime from Valhalla Wood Treatment
www.valhalco.com

PRODUCE DONATION

Plant a Row for the Hungry
https://gardencomm.org/PAR

HOME TEST KIT FOR ARSENIC

Industrial Test Systems, Inc.
www.sensafe.com/arsenic-kits

USEFUL APPS

Sun Seeker
Available on the Apple App Store and Google Play
Useful for figuring out sun exposure/shading on a potential garden space at any time of the year

Wild Edibles App
https://www.wildmanstevebrill.com/mobile-app

FROST DATES

Dave's Garden
https://davesgarden.com/guides/freeze-frost-dates

Farmer's Almanac Frost Dates Calculator
www.almanac.com/gardening/frostdates

MORE RESOURCES FROM SEATTLE URBAN FARM COMPANY

Seattle Urban Farm Company
www.seattleurbanfarmco.com
Visit our website to find downloadable versions of all the worksheets in this book, and to listen to our podcast, *Encyclopedia Botanica.*

METRIC CONVERSION CHARTS

Unless you have finely calibrated measuring equipment, conversions between US and metric measurements will be somewhat inexact. It's important to convert the measurements for all of the ingredients in a recipe to maintain the same proportions as the original.

VOLUME

TO CONVERT	TO	MULTIPLY
teaspoons	milliliters	teaspoons by 4.93
tablespoons	milliliters	tablespoons by 14.79
fluid ounces	milliliters	fluid ounces by 29.57
cups	milliliters	cups by 236.59
cups	liters	cups by 0.24
pints	milliliters	pints by 473.18
pints	liters	pints by 0.473
quarts	milliliters	quarts by 946.36
quarts	liters	quarts by 0.946
gallons	liters	gallons by 3.785

WEIGHT

TO CONVERT	TO	MULTIPLY
ounces	grams	ounces by 28.35
pounds	grams	pounds by 453.5
pounds	kilograms	pounds by 0.45

TEMPERATURE

TO CONVERT	TO	
Fahrenheit	Celsius	subtract 32 from Fahrenheit temperature, multiply by 5, then divide by 9

LENGTH

TO CONVERT	TO	MULTIPLY
inches	millimeters	inches by 25.4
inches	centimeters	inches by 2.54
inches	meters	inches by 0.0254
feet	meters	feet by 0.3048
feet	kilometers	feet by 0.0003048
yards	centimeters	yards by 91.44
yards	meters	yards by 0.9144
yards	kilometers	yards by 0.0009144
miles	meters	miles by 1,609.344
miles	kilometers	miles by 1.609344

INDEX

References to photos and illustrations are in italics.

extending the growing season
    with cold frames, 234–235
    with cool-weather crops, 229
    in fall and winter, 229, 235
    with greenhouses, 235–240
    hardy salad mix ingredients,
      230–231
    with heat-loving crops, 228
    with low tunnels, 232–235
    in spring and summer, 227–228

F

Fabaceae family, *51*, 52
farmers, thinking like, 12
fecal contamination, 251–252
fencing
    for pest control, 25, 156
    to support vining plants, 135
fennel, 50, 245, 249
fertility needs, 53
fertilizers
    blood meal, 139, 141
    foliar feeding, 140
    foliar feeds, 139
    general purpose mix, 139
    germination, 139
    high-phosphorus transplanting
      mix, 139
    labels, 140
    organic, 139, 141
    slow-release and soluble, 138–139
    sourcing, 64
    vegan, 139
    See also compost; manure; soil
      amendments
fertilizing
    crop-specific, 138, 142
    following transplanting, 122
    in the nursery, 194–196
    when to apply, 141
field peas, 91
    See also peas
fine seedbeds, 125
flame weeding, 98–99
    See also weed management
flats, 176, 201
floating row covers, 154–155, 227–
    228, 233, 235
    See also low tunnels
flowers
    anatomy of, 148
    discouraging blossoms, 145
    strategic removal of blossoms,
      144–145
    terminology, 148
    that attract beneficial insects, 154
fluorescent lights, 171
    See also lighting
foliar feeding, 139, 140

fresh crops, processing for storage,
    31
frost dates, 40, 42
frost-sensitive crops, 228
fungal inoculation, 117
fungus gnats, 198
    See also pest control

G

garlic
    'Chesnok Red', *27*
    crop rotation, 52
    curing for long storage, 261
    edible parts, 249
    'German White', *27*
    harvesting, 245, 255
germination fertilizer, 139
germination mix, 175, 201
germination rates, 177
gourds, 50
grafting, 191–192
grains, 31, 51
Granubor, 73
grass clippings, 94, 131
grass family, 51
greenhouses
    challenges of growing in, 237
    cover crops or fallow periods, 239
    coverings, 237
    and disease, 228
    finding the right greenhouse,
      236–237
    general information, 235–236
    heating, 174
    heat-loving crops in, 228
    inner covers, 240
    interior layout, 237–239
    laying out beds, 238
    lighting and heating, 239
    site selection and orientation, 237
    stand-alone, 169
    structure, 236
    thermometers, 238
    using to extend the season, 227–
      228, 239–240
    ventilation, 239–240
    See also home nursery
greens
    direct seeding, 124
    grouping crops, 186
    hardiness, 230
    nitrates in winter-harvested
      greens, 229
    relay planting, 36
    when to harvest, 255
    See also specific types
growing media, 174–176
    See also soil; soil amendments
gypsum, 72

H

hairy vetch, 52, 91
half-season crops, 28, 35
hand picking pests, 158
hand pollinating, 147
hardening off transplants, 196
hardwood cuttings, 188
harvest goals, 40
harvesting
    avoiding fecal contamination,
      251–252
    bolting plants, 144, 248
    cellular respiration, 253–254
    daily harvest timing, 254–255
    tools, 256–257
    washing and cooling, 255–256
    when to harvest, 244–247
harvest window, 109
heat cables, 174
heating
    in greenhouses, 239
    providing additional heat,
      173–174
    without electricity, 173
"heat island" effect, 23
heat-loving crops, 228
heat mats, 174
heat sinks, 23, 173
heavy metals, 72, 88
heirloom varieties, 107–108
herbaceous cuttings, 188
herbicides, 88
hermaphroditic plants, 147, 148
hidden crops, 248–250
HID lights, 172
    See also lighting
high-pressure sodium lights, 172
high tunnels. *See* greenhouses
hilling, 135
hoes, 98, 100–101
home nursery
    cleaning and storing trays and
      containers, 180
    containers, 176–179
    convenience, 168
    fertilizing plants, 194–196
    general information, 166–167
    growing media, 174–176
    growing microgreens, 199, 201,
      202–203
    growing sprouts, 199–200,
      202–203
    hardening off transplants, 196
    heating, 167, 173–174
    humidity, 167–168
    lighting, 167, 170–172
    locating the nursery, 169
    outfitting the nursery, 170–180
    pest and disease control, 196–198
    plant tags, 179–180

Garden Successfully with More Books from Storey

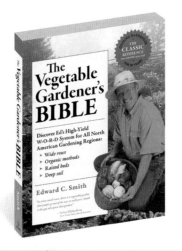

The Vegetable Gardener's Bible
by Edward C. Smith

Smith's legendary high-yield gardening methods result in bigger, better harvests. This comprehensive guide provides expert information and an inspiring roadmap for gardeners of all skill levels.

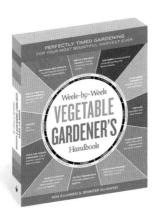

Week-by-Week Vegetable Gardener's Handbook
by Ron Kujawski & Jennifer Kujawski

Take the guesswork out of gardening! Detailed, customizable to-do lists break the work up into manageable tasks. This invaluable resource shows you exactly what to do—and exactly when and how to do it.

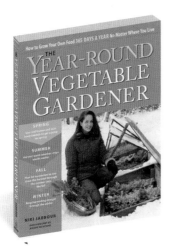

The Year-Round Vegetable Gardener
by Niki Jabbour

Grow your own food 365 days a year, no matter where you live. Learn the best varieties for each season, master succession planting, and make inexpensive protective structures that keep vegetables viable and delicious through the colder months.